CONSTRUCTION BUSINESS STARTUP 101

Laying the groundwork for success

Steven Smith

Wisdom Publishers

ISBN: 9798852504722
Imprint: Independently published

Cover design by: Art Painter
Library of Congress Control Number: 2018675309
Printed in the United States of America

To all the aspiring construction entrepreneurs,

This book is dedicated to your unwavering spirit, determination, and passion for building something extraordinary. May these pages empower you to turn your dreams into reality, navigate the challenges of entrepreneurship, and forge a path of success in the dynamic world of construction. Here's to the visionaries, innovators, and builders who dare to make a difference. May your journey be filled with growth, achievement, and the fulfillment of your entrepreneurial aspirations.

With heartfelt dedication,

[Steven Smith]

Every great dream begins with a dreamer. Always remember, you have within you the strength, the patience, and the passion to reach for the stars to change the world.

HARRIET TUBMAN

CONTENTS

INTRODUCTION

Welcome to this comprehensive guide on starting and succeeding as a construction entrepreneur. In the dynamic world of the construction industry, embarking on the journey of becoming a business owner requires a combination of industry knowledge, entrepreneurial skills, and unwavering determination. Whether you are a seasoned professional looking to launch your own construction venture or an aspiring entrepreneur seeking to make a mark in this competitive field, this guide is designed to equip you with the essential knowledge and encouragement needed to thrive.

The construction industry is an ever-evolving landscape, presenting a myriad of challenges and opportunities. This guide will walk you through the fundamental steps necessary to navigate this dynamic terrain successfully. From laying the groundwork with thorough market research and strategic planning to building a strong team, establishing a reputable brand, and harnessing the power of innovation, we will cover all aspects crucial for your construction business's success.

One of the key pillars of entrepreneurship is embracing continuous learning, and we will explore the significance of staying updated with industry trends, best practices, and emerging technologies. Throughout this journey, we will also draw inspiration from real-world stories of construction professionals who have triumphed over challenges, demonstrating the power of resilience and innovation in shaping their success.

Additionally, we will delve into the financial aspects of running a construction business, understanding budgeting, financial management, and risk assessment. Armed with this knowledge, you will be well-prepared to make informed decisions and optimize your business's financial stability and growth.

Networking and building relationships play a pivotal role in any entrepreneurial venture, and the construction industry is no exception. We will discuss the art of effective communication, building strategic partnerships, and leveraging your network to open doors to new opportunities.

Ultimately, this guide serves as a compass to steer you toward entrepreneurial greatness in the construction industry. Let us embark on this transformative journey together, exploring the world of construction entrepreneurship and discovering the untapped potential within you. With a spirit of innovation, a commitment to continuous growth, and a passion for building, you have the power to create a thriving construction business that leaves a lasting impact in this dynamic industry. Let's lay the foundation for your success and set forth on a path of growth, achievement, and fulfillment.

CHAPTER 1: INTRODUCTION TO STARTING A CONSTRUCTION BUSINESS

1.1 The Significance of the Construction Industry

The construction industry, often referred to as the backbone of modern civilization, holds immense significance in shaping our built environment and influencing our daily lives. From the infrastructure that connects communities to the architectural wonders that define skylines, construction plays a crucial role in fostering progress and development on a global scale.

At the heart of the construction industry lies its substantial contribution to economic growth and job creation. Governments and private entities invest heavily in construction projects, driving economic activity and providing employment opportunities to millions of people worldwide. As construction endeavors take shape, they create a ripple effect, stimulating demand for materials, skilled labor, and specialized services, benefitting not only the industry but also ancillary sectors that support its growth.

The construction industry has been at the forefront of innovation, continuously adapting to technological advancements to improve project delivery and efficiency. The incorporation of cutting-edge technologies such as Building Information Modeling (BIM), drones for aerial surveys, and artificial intelligence for project management has revolutionized the construction process. These advancements not only expedite construction timelines but also enhance precision and accuracy, reducing errors and rework.

Beyond its economic and technological impact, the construction industry plays a vital role in the creation of the built environment. Architectural marvels, engineered infrastructure, and functional spaces are the result of construction efforts that shape the fabric of our societies. From residential buildings that offer shelter and comfort to educational institutions that nurture young minds, construction projects serve the diverse needs of communities.

It is essential to understand that the construction industry significantly contributes to urbanization and regional development. Megacities are born as a result of vast construction projects that accommodate growing populations and drive urban expansion. In rural areas, construction plays a key role in providing essential infrastructure such as roads, bridges, and schools, fostering balanced development and access to opportunities.

The industry embraces sustainability and environmental responsibility. With a growing focus on green building practices, renewable energy installations, and eco-friendly materials, construction companies actively contribute to mitigating the impact of human activities on the planet. The integration of sustainable construction practices not only reduces resource consumption but also creates healthier and more resilient built environments.

The significance of this industry cannot be overstated. Its role in driving economic growth, job creation, and technological advancements is undeniable. The ability of the industry to shape the built environment and respond to the ever-changing needs of societies showcases its adaptability and forward-looking approach. As the world continues to evolve, the construction industry will remain at the forefront of progress, shaping the future we envision.

1.2 Challenges and Rewards of Starting a Construction Business

Starting a construction business can be a rewarding yet challenging endeavor. Aspiring entrepreneurs in the construction industry face a unique set of obstacles and opportunities as they embark on their journey. Understanding these challenges and rewards is crucial for those considering starting their own construction business.

One of the primary challenges of starting a construction business is securing sufficient financial resources. Acquiring necessary equipment, hiring skilled labor, and securing project sites require substantial upfront investment. In addition, construction projects often involve extended payment cycles, making effective cash flow management a critical aspect of the business.

Another significant challenge is navigating the complex landscape of regulatory compliance. The construction industry is subject to numerous regulations, permits, and licenses at various levels of government. Ensuring compliance with these requirements is essential to avoid penalties, project delays, and legal complications that can negatively impact the business.

Intense competition is a constant challenge in the construction industry. Established firms and experienced contractors often have a competitive advantage, making it difficult for new entrants to secure contracts. Building a reputation, demonstrating expertise, and differentiating the business from competitors are essential for winning contracts and gaining clients' trust.

Finding and retaining skilled labor is another ongoing challenge in the construction industry. Skilled workers are in high demand, and shortages can lead to project delays and increased labor costs. To attract and retain talented individuals, construction businesses must offer competitive compensation, provide opportunities for career development, and foster a positive work environment.

Managing project risks is a critical challenge in the construction industry. Construction projects are inherently prone to uncertainties such as weather delays, material shortages, and unforeseen site conditions. Effective risk management requires careful planning, proactive contingency measures, and open communication with all stakeholders involved.

Despite the challenges, starting a construction business offers numerous rewards. One of the most significant rewards is the opportunity for creative expression. Entrepreneurs in the construction industry have the chance to bring their visions to life, designing unique structures and implementing innovative construction techniques.

The tangible results of construction projects are also highly rewarding. Seeing a project come to fruition, from laying the foundation to the final touches, provides a deep sense of accomplishment and pride. Construction businesses have the privilege of leaving a lasting legacy through the structures they build, impacting communities for years to come.

Building a successful construction business also contributes to job creation and regional development. By providing employment opportunities, construction businesses play a vital role in supporting local economies. The construction of essential infrastructure and aesthetically pleasing spaces enhances communities and improves quality of life for residents.

Financial prosperity is another significant reward of starting a construction business. As the business grows and secures reputable projects, profit margins can increase, leading to financial stability and success. Entrepreneurs have the opportunity to build a profitable business while making a positive impact on their communities.

Personal growth is an inherent reward of starting and running a construction business. Entrepreneurs develop leadership skills, learn to navigate challenges, and make strategic decisions. The journey offers opportunities for self-improvement, resilience, and the satisfaction of overcoming obstacles.

Starting a construction business presents both challenges and rewards. Overcoming financial hurdles, navigating regulatory compliance, and managing project risks are critical to success. The rewards of creative expression, tangible results, job creation, financial prosperity, personal growth, and leaving a lasting impact on communities make the journey worthwhile. Entrepreneurs who embrace these challenges and persevere through obstacles are positioned to build successful and fulfilling construction businesses.

1.3 Key Factors for Success in the Construction Industry

Success in the construction industry hinges on various key factors that aspiring entrepreneurs and established companies must consider to thrive in this dynamic field. From strategic planning to cultivating a skilled workforce, the construction industry demands a multifaceted approach to achieve long-term success and sustain growth. Understanding and implementing these critical factors can position construction businesses for a competitive edge and pave the way for a flourishing future.

Strategic planning is at the heart of success in the construction industry. Developing a clear vision, mission, and long-term goals is essential to steer the business in the right direction. A well-defined strategy helps allocate resources effectively, identify potential opportunities and risks, and align the company's actions with its overarching objectives.

Market analysis and specialization play a crucial role in achieving success. Conducting comprehensive market research allows businesses to identify target markets, assess customer needs and preferences, and position themselves strategically within the industry. Specializing in specific areas of construction, such as residential, commercial, or industrial projects, can enable businesses to carve out a niche and differentiate themselves from competitors.

Building and maintaining strong relationships with clients are key factors for success. Providing exceptional customer service, delivering projects on time and within budget, and maintaining open lines of communication are vital. Satisfied clients not only bring repeat business but also serve as valuable references and sources of referrals.

A skilled and motivated workforce is an invaluable asset in the construction industry. Hiring and retaining qualified and experienced professionals, from project managers to skilled laborers, is essential for project success. Offering competitive compensation, ongoing training and development programs, and fostering a positive work culture can help attract and retain top talent.

Adapting to technological advancements is another critical factor for success. Embracing innovative construction technologies, such as Building Information Modeling (BIM), virtual reality, and drones, can enhance project efficiency, accuracy, and collaboration. Keeping pace with industry trends and leveraging technology to streamline processes and improve productivity can give construction businesses a competitive advantage.

Maintaining a strong focus on safety is paramount. Implementing robust safety protocols, providing adequate training, and promoting a safety-first culture are essential for protecting workers, reducing accidents, and mitigating liability. A commitment to safety not only safeguards the well-being of the workforce but also enhances the company's reputation and fosters client trust.

Effective project management is crucial for delivering projects successfully. Implementing efficient project management processes, from planning to execution and completion, helps ensure projects stay on schedule, within budget, and meet quality standards. Strong project management skills, effective communication, and proactive problem-solving are key to overcoming challenges and delivering exceptional results.

Adopting sustainable practices and incorporating green construction principles can position construction businesses for long-term success. Environmental consciousness is increasingly valued, and clients are prioritizing sustainability in their projects. Integrating sustainable design, construction materials, and energy-efficient technologies can not only reduce environmental impact but also attract environmentally conscious clients and demonstrate corporate social responsibility.

Continuous learning and improvement are integral to success in the construction industry. Staying updated on industry best practices, participating in professional development programs, and seeking opportunities for innovation and growth are crucial. Adapting to changing market demands and proactively evolving with industry trends ensure construction businesses remain relevant and competitive.

It is essential to understand that success in the construction industry requires a comprehensive approach that encompasses strategic planning, market analysis, client relationships, skilled workforce, technological adaptation, safety, project management, sustainability, and continuous improvement. By addressing these key factors, construction businesses can position themselves for growth, profitability, and long-term success in this dynamic and vital industry.

CHAPTER 2: RESEARCH AND PLANNING

2.1 Market Analysis: Understanding the Demand and Trends

Market analysis is a critical aspect of any business, and in the construction industry, it plays a pivotal role in understanding the demand and trends that influence project opportunities and business growth. By conducting comprehensive market analysis, construction companies can gain valuable insights into customer preferences, industry dynamics, and emerging trends. This knowledge empowers them to make informed decisions, tailor their services to meet market demands, and remain competitive in a constantly evolving landscape.

Understanding the demand is a fundamental component of market analysis. It involves identifying the types of projects that are in high demand, such as residential, commercial, industrial, or infrastructure developments. Additionally, assessing the overall construction activity, including the number and size of projects being undertaken, helps gauge the level of market activity.

Market analysis also involves studying customer preferences and requirements. This includes understanding the needs of different client segments, such as homeowners, businesses, government agencies, or developers. By identifying their preferences, construction companies can align their services and offerings to meet specific demands, whether it's sustainable construction practices, energy-efficient solutions, or innovative design concepts.

Another crucial aspect of market analysis is identifying target markets and evaluating their potential. This includes analyzing demographic factors such as population growth, income levels, and urbanization rates, as well as economic indicators that impact construction activity. By understanding the potential in different geographic areas, construction companies can prioritize their marketing and business development efforts accordingly.

In addition to understanding the demand, market analysis entails tracking industry trends. This includes staying abreast of technological advancements, emerging construction methods, and materials. Keeping an eye on regulatory changes, such as building codes and environmental standards, is also vital. By identifying these trends, construction companies can proactively adapt their practices, invest in necessary resources, and position themselves as industry leaders.

Market analysis involves competitor research to understand the competitive landscape. This includes identifying key competitors, their strengths, weaknesses, and market positioning. Analyzing their pricing strategies, service offerings, and customer satisfaction levels can provide valuable insights for developing competitive advantages and differentiation strategies.

Conducting a SWOT (Strengths, Weaknesses, Opportunities, and Threats) analysis is crucial. This assessment helps construction companies identify internal strengths and weaknesses, as well as external opportunities and threats. By understanding these factors, businesses can leverage their strengths, address weaknesses, capitalize on opportunities, and mitigate potential threats.

Market analysis is an ongoing process that requires continuous monitoring and evaluation. Construction companies should regularly review industry reports, attend trade shows and conferences, and engage with industry associations to stay informed about the latest developments and trends. Additionally, gathering feedback from customers, partners, and industry experts can provide valuable insights into evolving market demands.

Market analysis is a vital tool for construction companies to understand the demand and trends that shape the industry. By gaining a comprehensive understanding of customer preferences, target markets, industry trends, and competitive dynamics, construction businesses can make informed decisions, tailor their services to meet market demands, and position themselves for sustainable growth and success. Market analysis strategy enables construction companies to identify new opportunities, stay ahead of the competition, and build strong relationships with clients in an ever-changing industry.

2.2 Competitor Analysis: Identifying the Competition

Competitor analysis is a crucial component of business strategy, allowing construction companies to gain valuable insights into their competitors and make informed decisions to maintain a competitive edge. By conducting comprehensive competitor analysis, construction businesses can identify key players in the market, evaluate their strengths and weaknesses, understand their strategies, and identify opportunities for differentiation and growth. This knowledge empowers companies to develop effective strategies to outperform competitors and thrive in the construction industry.

Identifying the competition begins with conducting thorough research to identify key players in the construction market. This includes both direct competitors who offer similar services and solutions and indirect competitors who may provide alternative solutions to fulfill customer needs. By identifying the competition, construction companies can assess the level of market saturation and competition intensity.

Once the competitors are identified, it is essential to evaluate their strengths and weaknesses. This involves analyzing their areas of expertise, project portfolio, reputation, and customer base. By understanding their strengths, construction companies can gain insights into the factors that contribute to their success. On the other hand, assessing their weaknesses can help identify areas where the company can capitalize and differentiate itself.

Analyzing the strategies employed by competitors is another crucial aspect of competitor analysis. This includes evaluating their pricing strategies, marketing and advertising approaches, and customer engagement tactics. Understanding how competitors position themselves in the market and how they communicate their unique value proposition helps construction companies identify gaps and opportunities for differentiation.

Customer perception and satisfaction play a significant role in competitive analysis. Assessing customer reviews, testimonials, and feedback about competitors can provide insights into their reputation and level of customer satisfaction. By understanding customers' experiences with competitors, construction companies can identify areas for improvement and focus on delivering exceptional customer service to gain a competitive advantage.

Monitoring competitors' projects and recent successes can also provide valuable insights. Analyzing the types of projects they have secured, the scale of their operations, and their track record can help construction companies benchmark their own capabilities and identify opportunities for growth and expansion.

Keeping abreast of competitors' technological advancements and innovations is essential. This includes monitoring their adoption of new construction techniques, technologies, and tools. By understanding their technological investments, construction companies can assess the potential impact on project efficiency, quality, and client satisfaction. This knowledge enables businesses to make informed decisions about implementing new technologies within their own operations.

As applicable in market analysis, conducting a SWOT analysis for each competitor is essential. This analysis helps construction companies evaluate their competitors' strengths and weaknesses, as well as identify potential opportunities and threats they pose to the business. These factors can help construction companies develop strategies to capitalize on their competitors' weaknesses while mitigating potential threats.

It is important to note that competitor analysis is an ongoing process. Similar to market analysis, construction companies should continuously monitor their competitors' activities, stay updated on industry news and trends, and assess how their competitors adapt to changing market conditions. This allows businesses to proactively adjust their strategies, identify new opportunities, and maintain their competitive advantage.

Competitor analysis is an important tool for construction companies to stay competitive in the market. By identifying the competition, evaluating their strengths and weaknesses, analyzing their strategies, and monitoring their activities, construction companies can develop effective strategies for differentiation, capitalize on opportunities, and outperform competitors. A robust competitor analysis strategy allows businesses to position themselves strategically, deliver superior value to customers, and achieve long-term success in the construction industry.

2.3 Defining Your Target Audience and Ideal Clients

Defining your target audience and ideal clients is a crucial step in any business strategy, including in the construction industry. By identifying and understanding your target audience, construction companies can tailor their services, marketing efforts, and communication to effectively reach and engage potential clients. This process involves analyzing demographic factors, customer preferences, and market trends to create a clear picture of the ideal clients. By defining the target audience and ideal clients, construction companies can focus their resources and efforts on attracting and serving the most valuable customers in the market.

To define your target audience and ideal clients, you need to conduct thorough market research and analysis. Start by gathering demographic information such as age, gender, income level, and location. This data provides a foundational understanding of the characteristics of your potential customers and helps segment the market accordingly.

Once you have demographic insights, it's important to understand the psychographic factors that influence customer preferences and behaviors. This includes analyzing their values, interests, lifestyle choices, and purchasing habits. By delving deeper into these psychographic factors, construction companies can align their services and messaging to resonate with the target audience.

Market trends and industry dynamics play a significant role in defining the target audience and ideal clients. Stay updated on the latest trends and developments in the construction industry, such as sustainable construction practices, energy-efficient solutions, or smart technologies. Understanding these trends allows construction companies to position themselves strategically to cater to the evolving needs and preferences of the target audience.

Analyzing the pain points and challenges faced by potential clients is crucial. Identify the problems your target audience encounters in the construction process and assess how your services can address those issues effectively. By understanding their pain points, you can tailor your messaging to highlight the unique value and solutions your construction company offers.

Competitor analysis is also valuable in defining your target audience and ideal clients. Study the customer base of your competitors and identify gaps or underserved segments in the market. This analysis helps you identify opportunities to differentiate your services and create a competitive advantage in catering to specific customer needs and preferences.

Customer feedback and testimonials are excellent sources of information for defining the target audience and ideal clients. Engage with your existing clients to gather insights into why they chose your company, what they appreciate most about your services, and how you can further enhance their experience. Their feedback provides valuable insights into the characteristics and preferences of your ideal clients.

Developing buyer personas can be immensely helpful in defining the target audience and ideal clients. A buyer persona is a fictional representation of your ideal customer, incorporating various demographic, psychographic, and behavioral traits. Creating detailed buyer personas helps humanize your target audience, enabling you to understand their motivations, challenges, and decision-making processes.

Segmentation is a key element in defining the target audience and ideal clients. Divide the market into specific segments based on shared characteristics, such as project size, industry sector, or geographic location. By segmenting the market, you can tailor your marketing strategies and messages to resonate with each segment's unique needs and preferences.

Regularly review and refine your understanding of the target audience and ideal clients as market dynamics evolve. Stay updated on industry trends, conduct periodic market research, and seek feedback from your clients to ensure your strategies remain aligned with their changing needs.

Defining your target audience and ideal clients is a vital step in the success of a construction business. Through thorough market research, analyzing demographic and psychographic factors, considering industry trends, and understanding customer pain points, construction companies can create a clear picture of their target audience. This allows them to tailor their services, marketing efforts, and communication to effectively engage and serve their ideal clients. Focusing on the most valuable customers in the market, construction companies can position themselves for success, build strong client relationships, and achieve sustainable growth in the competitive construction industry.

2.4 Developing a Solid Business Plan for Your Construction Business

A good business plan is a foundational step for any construction business. A well-crafted business plan serves as a roadmap, guiding entrepreneurs through the intricacies of starting and running a successful construction company. It outlines the company's objectives, strategies, financial projections, and operational processes, providing a clear direction for achieving business goals. Crafting a comprehensive and realistic business plan requires careful consideration of market dynamics, competition, financial aspects, and operational requirements. By investing time and effort into creating a solid business plan, construction entrepreneurs set themselves up for a better chance of success in a competitive industry.

1. Executive Summary

The executive summary is a concise overview of the construction business plan, providing a snapshot of the company's mission, vision, key objectives, and competitive advantages. It highlights the market opportunity and outlines the main strategies the company will employ to achieve its goals.

2. Company Description

In the company description section, the construction business's background, legal structure, and ownership are detailed. This section should also include information about the services the company will offer, the target market, and any unique differentiators that set the business apart from competitors.

3. Market Analysis
The market analysis section delves into the construction industry's dynamics, including market size, growth trends, and customer segments. It analyzes the competitive landscape, identifies key competitors, and assesses their strengths and weaknesses. Market analysis provides insights into the target audience and helps define strategies for capturing market share.

4. Marketing and Sales Strategies
In this section, the construction company outlines its marketing and sales approaches. It specifies the marketing channels and tactics to reach the target audience effectively. The sales strategies detail how the company will acquire and retain customers, including pricing, promotions, and customer relationship management.

5. Services and Offerings
Here, the construction company describes its range of services and offerings in detail. It explains the scope of work, types of projects undertaken, and any specialty services provided. Clear descriptions help potential clients understand what the company can deliver and why they should choose its services.

6. Operational Plan
The operational plan outlines the day-to-day operations of the construction business. It includes information on staffing requirements, project management processes, quality control measures, and safety protocols. This section also highlights the company's facilities, equipment, and suppliers.

7. Financial Projections
Financial projections are a critical aspect of the business plan, providing insights into the company's expected revenues, expenses, and profitability over a specific period. It includes

projected income statements, balance sheets, and cash flow statements. Realistic financial projections help investors and lenders assess the company's potential for success.

8. Funding Requirements

For startups or businesses seeking funding, this section details the funding requirements to launch or expand the construction company. It outlines the amount of capital needed, how the funds will be utilized, and the potential sources of funding, such as equity investment or loans.

9. Risk Analysis and Mitigation

The risk analysis section identifies potential risks and challenges that may affect the construction business's success. It outlines strategies to mitigate these risks and contingency plans for unforeseen circumstances. Demonstrating a proactive approach to risk management boosts confidence in the company's ability to navigate challenges.

10. Conclusion

The conclusion summarizes the key points of the business plan and reiterates the company's vision and objectives. It provides a compelling case for why the construction business is poised for success and sets a positive tone for potential investors, partners, and stakeholders.

Developing a solid business plan is crucial for the success of a construction business. A well-structured business plan serves as a roadmap, guiding the company through the complexities of the construction industry and positioning it for sustainable growth and profitability. Through careful analysis of market dynamics, competition, operational requirements, and financial projections, construction entrepreneurs can create a compelling business plan that inspires confidence in potential investors and sets the stage for a successful venture in the competitive construction landscape.

CHAPTER 3: LEGAL AND REGULATORY CONSIDERATIONS

3.1 Legal Structure: Choosing the Right Business Entity

Selecting the appropriate legal structure is a critical decision for any construction business. The legal structure chosen determines the company's liability, tax obligations, management, and overall governance. Different legal entities offer varying levels of protection, tax benefits, and operational flexibility. Thus, it is essential for construction entrepreneurs to carefully evaluate their options and choose the business entity that aligns best with their long-term goals and business needs.

1. Sole Proprietorship
A sole proprietorship is the simplest and most common legal structure for small businesses, including construction companies. In this structure, the business is owned and operated by a single individual. While it is easy and inexpensive to set up, a sole proprietorship does not provide liability protection. The owner assumes full personal liability for business debts and legal obligations, which could put personal assets at risk.

2. Partnership
A partnership is formed when two or more individuals join together to run a construction business. Like sole proprietorships,

partnerships offer simplicity in formation. However, partners share both profits and liabilities, making them jointly responsible for the business's debts and obligations. It is essential for partners to have a clear and well-drafted partnership agreement to outline ownership stakes, decision-making authority, and profit-sharing arrangements.

3. Limited Liability Company (LLC)
A LLC is a popular choice for many construction businesses due to its flexibility and liability protection. In an LLC, owners, known as members, are not personally liable for the company's debts and legal obligations. This legal structure combines features of both partnerships and corporations, providing pass-through taxation and a simplified management structure. LLCs offer the benefit of protecting personal assets while maintaining relative ease of operation.

4. Corporation
A corporation is a separate legal entity from its owners, known as shareholders. This legal structure offers the highest level of liability protection, as shareholders are generally not personally liable for the company's debts and liabilities. Corporations have a formal management structure, including a board of directors and officers. They may be subject to double taxation, with the company being taxed on its profits, and shareholders facing taxation on dividends. The formation and maintenance of a corporation involve more extensive legal and administrative requirements.

5. S Corporation
An S corporation is a special type of corporation that allows for pass-through taxation. It avoids double taxation by passing profits and losses directly to shareholders' personal tax returns. To qualify as an S corporation, the business must meet specific IRS requirements, such as having a limited number of shareholders and being a domestic corporation. The S corporation structure

offers liability protection while providing tax advantages similar to those of an LLC.

6. Professional Corporation (PC) or Professional Limited Liability Company (PLLC)

For construction businesses that involve licensed professionals, such as architects or engineers, a professional corporation (PC) or professional limited liability company (PLLC) may be suitable. These entities offer liability protection for the individual professionals while still allowing them to provide professional services in their respective fields.

Choosing the right legal structure for a construction business depends on various factors, including the level of liability protection needed, tax implications, management preferences, and long-term goals. Here are some key considerations to help construction entrepreneurs make an informed decision:

1. Liability Protection
Protecting personal assets from business debts and legal liabilities is a primary concern for construction entrepreneurs. Legal entities such as LLCs and corporations offer limited liability, shielding owners from personal responsibility for the company's obligations.

2. Tax Implications
Different legal structures have distinct tax implications. Sole proprietorships, partnerships, and S corporations offer pass-through taxation, where business profits and losses flow directly to the owners' personal tax returns. Corporations, on the other hand, face corporate-level taxation, which may result in double taxation when distributing profits to shareholders. The specific tax benefits and obligations must be carefully assessed in light of the company's financial situation and long-term plans.

3. Management Structure
The desired management structure of the construction business

is another essential factor. Sole proprietorships and partnerships offer simplicity in decision-making and management, as the owner(s) have direct control. Corporations and LLCs typically have more formalized management structures, with clearly defined roles for directors, officers, and members.

4. Business Continuity

Considerations regarding business continuity and succession planning are crucial for the long-term success of the construction business. Legal entities like corporations offer more straightforward procedures for transferring ownership or raising capital through the issuance of stock.

5. Administrative Requirements

The administrative requirements for each legal structure vary. Sole proprietorships and partnerships have minimal formalities, while corporations and LLCs often require more paperwork and compliance with state regulations. Understanding the administrative burden can help construction entrepreneurs manage their time and resources efficiently.

6. Future Growth and Investment

When planning for the future growth of the construction business, entrepreneurs must consider how their chosen legal structure may impact raising capital and attracting investors. Corporations, for example, have a more established framework for issuing shares and attracting shareholders.

7. Legal and Professional Advice

Given the complexity of legal structures and their implications, seeking legal and professional advice is highly advisable. Consulting with a business attorney and an accountant with experience in the construction industry can provide invaluable insights and ensure compliance with all legal requirements.

Choosing the right legal structure is a significant decision that lays the foundation for a construction business's success. The legal structure selected affects liability, taxation, management, and overall governance. Construction entrepreneurs must carefully assess their needs, long-term goals, and risk tolerance to determine the most suitable legal entity. Seeking professional guidance ensures compliance with legal requirements and positions the construction business for a stable and prosperous future in a competitive industry.

3.2 Licensing and Permits: Meeting Regulatory Requirements

Meeting regulatory requirements for licensing and permits is a crucial aspect of operating a construction business. The construction industry is heavily regulated to ensure safety, quality, and compliance with environmental and building standards. Obtaining the necessary licenses and permits is not only a legal obligation but also a fundamental step toward building a reputable and successful construction business. This article explores the significance of licensing and permits in the construction industry, the types of licenses and permits required, and the process of obtaining them.

Understanding the Significance

Licensing and permits are essential in the construction industry to safeguard public safety, protect the environment, and maintain quality standards. Construction projects involve significant risks and potential hazards, making adherence to regulations paramount to prevent accidents and ensure proper construction practices. Furthermore, licensing and permits provide assurance to clients that the construction company possesses the necessary qualifications and approvals to carry out the proposed work competently.

Types of Licenses and Permits

The specific licenses and permits required for a construction business can vary depending on the region, type of construction work, and project scope. Some common types of licenses and permits include:

Business License: A general business license is typically required to operate any type of business, including construction companies. It validates the legality of the business and allows it to conduct operations in the specified jurisdiction.

Contractor's License: In many jurisdictions, construction contractors are required to obtain a contractor's license. This license verifies that the contractor has the necessary skills and expertise to perform construction work and ensures compliance with local building codes.

Building Permits: Building permits are essential for construction projects and are typically issued by local authorities. They authorize specific construction activities and confirm that the plans comply with building codes and regulations.

Environmental Permits: Certain construction projects may require environmental permits, especially if they involve activities that could impact the environment, such as excavation near water bodies or construction in protected areas.

Safety Certifications: Construction companies may need safety certifications to demonstrate that they adhere to safety standards and protocols. These certifications are particularly crucial in hazardous construction environments.

Zoning and Land Use Permits: Before commencing construction, companies may need to obtain zoning and land use permits to ensure that the proposed project aligns with the designated land use and zoning regulations.

Occupational Licenses: In some cases, individuals involved in specialized construction trades, such as electricians or plumbers, may require specific occupational licenses to practice legally.

Obtaining Licenses and Permits

The process of obtaining licenses and permits can be complex and time-consuming. It typically involves the following steps:

Research: Construction companies should start by researching the specific licenses and permits required for their type of work and location. This information can be obtained from local government websites or regulatory agencies.

Application: After identifying the required licenses and permits, companies need to complete the necessary application forms accurately and provide any supporting documentation as requested.

Fees and Payments: Licensing and permit applications often involve associated fees. Companies must ensure they pay the required fees promptly to avoid delays in processing.

Inspections and Reviews: Depending on the type of permit, inspections and reviews may be required before approval is granted. This process ensures that the construction plans meet safety and quality standards.

Renewal and Compliance: Many licenses and permits have expiration dates and require periodic renewal. Companies must keep track of renewal dates and comply with any ongoing requirements to maintain their valid status.

Consultation and Legal Support: Given the complexity of licensing and permitting processes, seeking consultation from legal experts or construction consultants can be beneficial. These professionals can provide guidance on navigating the regulatory landscape and ensure compliance with all requirements.

Benefits of Compliance

Complying with licensing and permitting requirements offers numerous benefits to construction businesses:

Legal Protection: Proper licensing and permitting protect construction companies from legal liabilities and potential fines resulting from non-compliance with regulations.

Client Trust: Clients are more likely to trust a licensed and permitted construction company, as it demonstrates the company's commitment to adhering to industry standards and delivering quality work.

Competitive Advantage: Having the necessary licenses and permits can provide a competitive edge, as it distinguishes the company as a reputable and professional entity within the industry.

Expanded Opportunities: Compliance with regulatory requirements opens up opportunities for bidding on government contracts and participating in larger and more complex construction projects.

Safety and Quality Assurance: Following licensing and permitting requirements ensures that construction work is carried out safely and to the required quality standards.

Obtaining the appropriate licenses and permits is a fundamental requirement for operating a construction business. It not only ensures legal compliance but also serves as a mark of professionalism and commitment to safety and quality standards. Understanding the types of licenses and permits needed, the application process, and the benefits of compliance is crucial for construction entrepreneurs to navigate the regulatory landscape successfully. By proactively meeting regulatory requirements, construction companies can establish themselves as reliable and reputable players in the industry, gaining the trust of clients and positioning themselves for long-term success and growth.

3.3 Zoning and Building Codes: Understanding Local Regulations

Understanding local zoning and building codes is essential for any construction project. Zoning and building codes are regulations imposed by local governments to control land use and ensure that construction projects meet specific safety and quality standards. These regulations vary from one jurisdiction to another and can significantly impact the planning, design, and execution of construction projects. Complying with zoning and building codes is not only a legal requirement but also critical for the successful completion of construction projects and for avoiding potential penalties and delays.

Zoning Regulations

Zoning regulations divide land into different zones or districts, each with specific permitted uses and restrictions. These regulations are designed to promote orderly development, protect property values, and prevent incompatible land uses. Common zoning classifications include residential, commercial, industrial, agricultural, and mixed-use zones. Understanding the zoning regulations in a particular area is vital for construction companies and developers to determine whether their proposed projects align with the designated land use.

1. Zoning Maps and Ordinances
Local governments provide zoning maps and ordinances that outline the different zones and their permitted uses. These resources help construction professionals identify the zoning designation of a specific property and assess the types of construction projects allowed in that area.

2. Special Use Permits and Variances
In some cases, construction projects may not conform to the strict zoning regulations. However, they may still be permitted through special use permits or variances. Special use permits are granted for projects that serve a public benefit but may not be consistent with the zoning code. Variances, on the other hand, allow for

minor deviations from zoning requirements based on unique circumstances.

Building Codes

Building codes are a set of regulations that govern the design, construction, and occupancy of buildings to ensure the safety and well-being of occupants. Building codes cover various aspects of construction, including structural integrity, fire safety, electrical systems, plumbing, and accessibility. Adhering to building codes is critical to constructing safe and habitable buildings that meet the required standards.

1. International Building Code (IBC)
The International Building Code is a widely adopted model code that sets the minimum requirements for building construction and design in the United States and many other countries. Local governments often adopt the IBC as the basis for their own building codes, with additional regional amendments as needed.

2. Local Amendments
Many jurisdictions have specific local amendments to the building code to address unique regional concerns, climate conditions, or geological factors. Construction professionals must familiarize themselves with these local amendments to ensure compliance.

3. Inspections and Permits
To ensure compliance with building codes, construction projects typically require various inspections at different stages of construction. Obtaining the necessary building permits is a critical step in the construction process, as it confirms that the project plans meet the applicable building codes.

Challenges and Considerations

Navigating local zoning and building codes can present challenges for construction professionals:

1. Complex and Changing Regulations
Zoning and building codes can be complex and subject to frequent updates. Keeping up with the latest regulations and amendments requires diligence and attention to detail.

2. Compliance with Multiple Jurisdictions
For construction companies operating in multiple jurisdictions, compliance with different zoning and building codes can be challenging. Each local government may have its own set of rules and requirements.

3. Time and Cost Impact
Complying with zoning and building codes may impact project timelines and costs. Special use permits, variances, and additional inspections can cause delays and increase expenses.

4. Design Flexibility
Zoning regulations may limit the design flexibility of a project, especially in areas with strict design standards and building height restrictions.

Benefits of Compliance

Complying with local zoning and building codes offers several benefits:

1. Legal Compliance
Adhering to zoning and building codes is a legal requirement. Failure to comply can result in fines, penalties, or even the suspension of construction activities.

2. Safety and Quality Assurance
Building codes are designed to ensure the safety and structural integrity of buildings. Complying with these codes results in safer and higher-quality construction projects.

3. Community Acceptance
Compliance with zoning regulations fosters positive relationships with the community and local authorities, enhancing the

acceptance of construction projects.

4. Streamlined Approval Process
Projects that meet zoning and building code requirements are more likely to receive timely approvals and permits, expediting the construction process.

Understanding local zoning and building codes is a fundamental aspect of successful construction projects. Zoning regulations dictate land use and permitted construction activities, while building codes ensure safety and quality standards. Construction companies must be diligent in researching and complying with these regulations to avoid legal issues, ensure safe and reliable construction, and foster positive relationships with the community and local authorities. By prioritizing compliance with zoning and building codes, construction professionals can navigate the regulatory landscape effectively and achieve successful project outcomes.

3.4 Insurance Coverage: Protecting Your Business and Employees

Insurance is a vital aspect of risk management for construction businesses. It provides financial protection against various risks and liabilities that are inherent in the construction industry. From safeguarding assets and projects to protecting employees and clients, insurance coverage plays a crucial role in ensuring the stability and success of a construction business.

General liability insurance is a fundamental coverage that all construction companies should have. It protects businesses from third-party claims for property damage, bodily injury, and personal injury that may arise during construction projects. For example, if a construction worker accidentally damages a client's property, general liability insurance would cover the costs of repair or replacement.

Workers' compensation insurance is a mandatory coverage in

most jurisdictions for businesses with employees. It provides coverage for medical expenses, lost wages, and disability benefits to employees who sustain work-related injuries or illnesses. Workers' compensation insurance not only protects employees but also shields employers from potential lawsuits related to workplace injuries.

Commercial property insurance is essential for protecting a construction company's physical assets, such as office buildings, equipment, tools, and machinery. It provides coverage for damages caused by events like fire, theft, vandalism, or natural disasters. With construction projects often involving valuable equipment and materials, having commercial property insurance is critical to recover from potential losses.

Professional liability insurance, also known as errors and omissions (E&O) insurance, is crucial for construction companies that offer professional services. It protects against claims of negligence, errors, or omissions in the services provided. For example, if a construction company provides flawed design plans that lead to project delays and financial losses for the client, professional liability insurance would cover the legal costs and potential damages.

Commercial auto insurance is necessary for construction businesses that use vehicles for transportation, such as trucks, vans, or cars. This coverage protects against accidents, damages, and injuries involving company-owned vehicles. Whether it's transporting materials to a construction site or traveling to meet clients, commercial auto insurance provides financial protection in case of unforeseen incidents.

Builder's risk insurance, also known as course of construction insurance, is specific to construction projects. It provides coverage for damages to buildings or structures under construction. It protects against risks like theft, fire, vandalism, or weather-related damage until the project is completed and handed over to the client.

In addition to the standard insurance coverages mentioned above, construction companies may require additional specialized policies based on their specific operations and project requirements. These additional coverages might include pollution liability insurance, subcontractor insurance, equipment breakdown insurance, and surety bonds, among others.

When determining the appropriate insurance coverage for a construction business, it is essential to assess the specific risks associated with the company's operations, the size of the projects undertaken, the number of employees, and the local regulatory requirements. Working with an experienced insurance broker or agent can help construction companies tailor insurance policies to their unique needs and ensure they have comprehensive coverage to protect their assets, projects, employees, and reputation.

Having adequate insurance coverage not only safeguards a construction business from financial losses but also instills confidence in clients and partners. Clients are more likely to choose construction companies with robust insurance coverage, as it demonstrates the company's commitment to risk management and responsible business practices.

Insurance coverage is a crucial investment for construction businesses to protect against various risks and liabilities. General liability insurance, workers' compensation insurance, commercial property insurance, professional liability insurance, and commercial auto insurance are among the essential coverages that construction companies should have. Additionally, builder's risk insurance and other specialized policies can provide specific project-related protection. By carefully assessing their risk exposure and working with insurance professionals, construction companies can ensure they have comprehensive coverage to secure their assets, projects, and employees, allowing them to focus on delivering successful construction projects with confidence and peace of mind.

CHAPTER 4: FINANCIAL PREPARATION

4.1 Estimating Startup Costs and Budgeting

One of the significant startup costs for a construction business is acquiring the necessary equipment and tools. Depending on the type of construction projects the company intends to undertake, this may include excavators, bulldozers, cranes, trucks, scaffolding, power tools, and safety equipment. Evaluating the market for both new and used equipment can help determine the best options to balance quality and affordability.

The cost of materials and supplies is another critical consideration for construction businesses. Whether it's concrete, steel, lumber, or other building materials, accurately estimating the quantity and cost of materials required for initial projects is essential for budgeting. Building relationships with suppliers and negotiating favorable terms can also help reduce procurement costs in the long run.

Labor costs represent a significant portion of a construction business's expenses. Depending on the size and scope of projects, construction companies may need to hire skilled laborers, project managers, engineers, and administrative staff. Recruiting and retaining qualified employees is crucial for the success of construction projects. While experienced workers may demand higher salaries, investing in skilled labor ensures the delivery of high-quality projects.

Renting or purchasing office space and construction yards is a significant expense for construction companies. The location and size of these facilities should be carefully considered to accommodate the company's operations and future growth. Additionally, utilities, insurance, and other overhead costs associated with the office and yard need to be factored into the budget.

Insurance coverage is an essential aspect of financial planning for a construction business. As discussed earlier, insurance protects the company against various risks and liabilities, including property damage, workplace injuries, and professional errors. Premiums for general liability insurance, workers' compensation insurance, and other coverage must be accounted for in the startup budget.

Legal and permit fees are often overlooked but can quickly add up during the startup phase. Construction companies must obtain the necessary licenses and permits to operate legally and undertake construction projects in compliance with local regulations. Understanding the specific requirements and costs for permits, inspections, and other legal obligations is vital for accurate budgeting.

Marketing and advertising expenses are necessary to promote the construction business and attract clients. Developing a professional website, creating marketing materials, attending industry events, and networking with potential clients all incur costs. Allocating a budget for marketing activities helps the company establish its brand and generate leads.

Contingency funds are a critical component of the budgeting process. Construction projects are inherently subject to unforeseen challenges, such as weather delays, supply chain disruptions, or design changes. Setting aside a contingency reserve helps mitigate the impact of these unforeseen events and prevents budget overruns.

Tracking expenses and revenues is essential for financial management. Implementing robust accounting software and hiring a qualified accountant or bookkeeper can ensure accurate record-keeping and financial reporting. Regular financial reviews and analysis provide insights into the company's financial performance, enabling timely adjustments to the budget and operations.

As the construction business progresses, ongoing budgeting remains vital for maintaining financial health. Regularly reviewing the budget and comparing actual expenses against projections helps identify areas for improvement and cost-saving opportunities. A well-managed budget allows the construction business to stay financially stable, fulfill its financial obligations, and make strategic decisions for growth and expansion.

Estimating startup costs and budgeting are crucial steps for any construction business. Careful analysis and planning ensure that the company has sufficient funds to cover initial expenses, purchase necessary equipment and materials, and recruit skilled labor. Budgeting also accounts for ongoing operational costs, including office space, insurance, permits, marketing, and contingency reserves. Regularly tracking expenses and revenues allows the construction business to make informed financial decisions, stay on track with financial goals, and adapt to changing market conditions. With a well-prepared budget, construction companies can confidently embark on their projects, secure in the knowledge that their financial resources are allocated efficiently for long-term success.

4.2 Funding Options for Your Construction Business

Funding options play a crucial role in the success and growth of a construction business. Whether it's starting a new venture or expanding existing operations, having access to adequate capital is essential to cover startup costs, purchase equipment, hire skilled labor, and support ongoing operations. Understanding the various funding options available can help construction entrepreneurs make informed decisions and secure the necessary financing to fuel their business growth.

One of the primary funding options for construction businesses is traditional bank loans. These loans typically require a detailed business plan, financial projections, collateral, and a good credit history. Bank loans provide a lump sum amount that can be used to cover startup costs, purchase equipment, or invest in infrastructure. Repayment terms, interest rates, and eligibility criteria vary depending on the bank and the specific loan program.

Another funding option for construction businesses is Small Business Administration (SBA) loans. The SBA provides loan programs specifically designed to assist small businesses, including those in the construction industry. SBA loans offer competitive interest rates, longer repayment terms, and flexible requirements. However, the application process may be more rigorous, requiring thorough documentation and compliance with SBA guidelines.

Construction businesses can also explore equipment financing options. Equipment financing allows companies to acquire necessary machinery, tools, and vehicles without making a significant upfront payment. The equipment itself serves as collateral for the loan, and the repayment terms are often tied to the useful life of the equipment. This type of financing helps conserve cash flow while ensuring access to the necessary equipment for construction projects.

Invoice financing, also known as accounts receivable financing, is a funding option that enables construction businesses to access immediate cash flow by selling their outstanding invoices to a financing company. This can be particularly useful for businesses that experience delays in receiving payments from clients. Invoice financing provides a percentage of the invoice amount upfront, with the remaining balance minus a financing fee paid once the client settles the invoice.

Construction companies can also consider lines of credit as a flexible funding option. A line of credit allows businesses to access a predetermined amount of funds as needed. Interest is only charged on the funds utilized, providing greater flexibility and cost control. Lines of credit can be used to cover short-term working capital needs, such as purchasing materials, meeting payroll, or handling unexpected expenses.

Private investors or venture capitalists can be a viable funding option for construction businesses with high growth potential. These investors provide capital in exchange for equity or a share of the company's profits. Partnering with an investor brings not only financial resources but also valuable industry expertise and connections. However, it's important to carefully consider the terms and implications of such partnerships before entering into any agreements.

Crowdfunding has emerged as an alternative funding option for businesses across various industries, including construction. Through online platforms, construction companies can raise funds by soliciting small contributions from a large number of individuals. Crowdfunding allows businesses to engage with potential clients and supporters while generating the necessary capital for specific projects or business expansion.

Government grants and subsidies may be available for construction businesses that meet specific criteria. These grants can provide financial support for research and development, energy-efficient construction, or projects that benefit the community. However, obtaining government grants often requires a competitive application process and adherence to specific guidelines.

Funding options for construction businesses are diverse and can be tailored to specific needs and circumstances. Traditional bank loans, SBA loans, equipment financing, invoice financing, lines of credit, private investors, crowdfunding, and government grants are among the funding options available. Choosing the most suitable option requires a thorough assessment of the business's financial requirements, growth plans, and risk tolerance. Additionally, understanding the terms, repayment structures, and associated costs is crucial in making informed decisions. Seeking professional advice from financial advisors or consultants with expertise in the construction industry can provide valuable insights and guidance throughout the funding process. By securing the necessary capital, construction businesses can pursue growth opportunities, deliver high-quality projects, and thrive in a competitive market.

4.3 Creating Financial Projections and Forecasts

Creating financial projections and forecasts is an essential aspect of financial planning for a construction business. These projections provide a clear and realistic picture of the company's financial future, allowing entrepreneurs to make informed decisions, set financial goals, and secure funding from investors or lenders. By analyzing past financial data and considering market trends, construction businesses can develop accurate projections that guide their strategic planning and operational decisions.

One of the key elements of financial projections is the income statement, also known as the profit and loss (P&L) statement. The income statement projects the company's revenues, expenses, and net income over a specific period, typically one year. It includes details on revenues generated from construction projects, operating expenses, salaries, and other costs incurred during the period. By forecasting revenues and expenses, construction businesses can estimate their profitability and identify areas for cost management and efficiency improvement.

The balance sheet is another essential component of financial projections. The balance sheet provides a snapshot of the company's financial position at a specific point in time, listing its assets, liabilities, and shareholders' equity. Assets include cash, accounts receivable, equipment, and property, while liabilities encompass accounts payable, loans, and other debts. By projecting the company's assets and liabilities, construction businesses can assess their liquidity, financial health, and capacity to take on additional debt or investments.

Cash flow projections are vital for construction companies, as managing cash flow is crucial to sustaining day-to-day operations. Cash flow projections track the inflow and outflow of cash over a specified period, such as a month or a quarter. By forecasting cash flow, construction businesses can anticipate times of potential cash shortages and take proactive measures to ensure they have sufficient funds to meet their financial obligations.

To develop accurate financial projections, construction businesses must consider several factors. Market trends, economic conditions, and industry growth rates can significantly impact revenue forecasts. Analyzing past financial performance and identifying patterns and trends can provide valuable insights into the company's financial behavior and help develop realistic projections.

Estimating project costs and revenues is critical for construction companies. Assessing the potential profitability of specific projects enables businesses to prioritize and allocate resources effectively. This involves analyzing the costs of labor, materials, equipment, and subcontractors, as well as estimating the revenue potential of each project.

Assumptions play a key role in financial projections. Construction businesses must make reasonable assumptions about various aspects of their operations, such as sales growth, inflation rates, and interest rates. These assumptions should be based on industry research, market data, and historical performance.

Sensitivity analysis is a useful tool to assess the impact of different scenarios on financial projections. By varying key assumptions, construction businesses can understand how changes in market conditions or internal factors may affect their financial performance. This allows them to make more informed decisions and develop contingency plans to navigate potential challenges.

Financial projections should be regularly reviewed and updated to reflect changes in the business environment and performance. As construction projects are often subject to uncertainties, revisions to financial projections help maintain accuracy and relevance. Regularly comparing actual financial performance to projections also provides valuable feedback on the accuracy of assumptions and the effectiveness of financial planning.

Creating accurate and realistic financial projections is essential for construction businesses to make informed financial decisions, secure funding, and plan for future growth. The income statement, balance sheet, and cash flow projections are vital components of financial projections that provide insights into the company's revenue, expenses, profitability, and liquidity. Accurate forecasting involves analyzing market trends, project costs, and historical performance, while making reasonable assumptions and conducting sensitivity analysis. Regularly updating financial projections and comparing actual performance to projections help construction businesses adapt to changing market conditions and maintain financial stability. By developing well-informed financial projections, construction businesses can navigate uncertainties, make strategic decisions, and achieve their long-term financial goals.

4.4 Managing Cash Flow Effectively

Managing cash flow effectively is crucial for the success and sustainability of any business, including construction companies. Cash flow refers to the movement of money in and out of a business over a specific period, such as a month or a quarter. Positive cash flow occurs when a business generates more cash than it spends, while negative cash flow indicates that more cash is flowing out than coming in. Proper cash flow management is essential for meeting financial obligations, paying suppliers and employees, investing in growth opportunities, and maintaining a healthy financial position.

One of the first steps in managing cash flow effectively is to create a detailed cash flow forecast. A cash flow forecast estimates the expected inflows and outflows of cash over a specific period. It helps construction businesses anticipate periods of cash surplus or shortfall, allowing them to plan accordingly. The forecast should include revenue projections, operating expenses, loan repayments, and any other cash-related transactions.

One effective strategy for managing cash flow is to implement a system for invoicing and receivables management. Construction projects often involve billing clients in multiple stages, such as upon signing the contract, reaching project milestones, and upon completion. Timely and accurate invoicing helps ensure that clients are aware of their payment obligations and reduces the risk of delays in receiving payments.

To improve cash flow, construction businesses should consider offering discounts or incentives for early payment. Providing clients with a small percentage discount for paying invoices before the due date can incentivize prompt payment and help maintain a positive cash flow.

Efficient inventory management is essential for construction companies to avoid tying up excessive capital in materials and supplies. Overstocking can lead to increased storage costs and tie up cash that could be better used elsewhere. Regularly reviewing inventory levels and optimizing the procurement process can help reduce carrying costs and improve cash flow.

Negotiating favorable payment terms with suppliers can also positively impact cash flow. Construction businesses should strive to extend payment terms with suppliers while maintaining good relationships. By having more time to pay for materials and services, businesses can better align their cash outflows with cash inflows from client payments.

Construction companies should regularly review their expenses to identify potential areas for cost savings. Trimming unnecessary expenses and renegotiating contracts with vendors can free up cash and improve overall cash flow.

For construction businesses that experience seasonality or periodic fluctuations in revenue, creating a cash reserve can help bridge the gap during lean periods. Setting aside a portion of profits during high-revenue months can provide a safety net during slower times.

Effective cash flow management also involves closely monitoring accounts receivable and managing collections proactively. Delayed or non-payment from clients can significantly impact cash flow and hinder a company's ability to meet financial obligations. Implementing a robust collection process, including regular follow-ups with clients, can help improve cash flow by reducing outstanding receivables.

Construction businesses should also maintain a clear and accurate picture of their cash flow through regular financial reporting and analysis. Utilizing accounting software or engaging with financial professionals can aid in monitoring cash flow, identifying trends, and making informed decisions based on financial data.

Another strategy for improving cash flow is to negotiate payment terms with clients. When entering into contracts, construction companies can negotiate shorter payment cycles to receive payments sooner and maintain steady cash flow.

Managing cash flow effectively is vital for the financial health and success of construction businesses. Creating a detailed cash flow forecast, implementing efficient invoicing and receivables management systems, and optimizing inventory and expenses are essential strategies for improving cash flow. Offering incentives for early payment and negotiating favorable payment terms with suppliers can also positively impact cash flow. Regularly monitoring cash flow through financial reporting and analysis allows construction businesses to make informed decisions and maintain financial stability. By employing these strategies and maintaining a proactive approach to cash flow management, construction companies can navigate economic fluctuations, meet financial obligations, and position themselves for long-term success.

CHAPTER 5: BUILDING A STRONG TEAM

5.1 Identifying the Key Roles for Your Construction Business

Identifying the key roles for a construction business is a crucial step in building a strong and efficient team that can deliver successful projects and drive business growth. Each role within the construction company plays a unique and essential part in ensuring smooth operations, effective project management, and overall success. From leadership positions to skilled trades and support roles, understanding the responsibilities and qualifications for each key role is essential for recruiting the right talent and fostering a collaborative and productive work environment.

Leadership Roles

Chief Executive Officer (CEO) / President: The CEO or President is the top executive responsible for setting the overall vision, strategy, and direction of the construction business. They make high-level decisions, oversee financial performance, and are the main point of contact for stakeholders, clients, and partners.

Chief Operations Officer (COO) / Vice President (VP) of

Operations: The COO or VP of Operations is in charge of the day-to-day management of the construction company. They ensure that projects are executed efficiently, monitor project progress, and implement processes to enhance productivity.

Chief Financial Officer (CFO) / Finance Director: The CFO or Finance Director oversees the financial aspects of the construction business, including budgeting, financial planning, and financial reporting. They ensure that the company's financial resources are effectively managed and allocate funds to support growth initiatives.

Project Manager: Project managers are responsible for overseeing individual construction projects from start to finish. They coordinate teams, manage resources, ensure compliance with timelines and budgets, and act as the primary point of contact for clients.

Skilled Trades

Construction Foreman: The construction foreman is a supervisory role responsible for leading the on-site construction team. They ensure that construction activities are carried out safely, efficiently, and according to plans and specifications.

Carpenter: Carpenters are skilled tradespeople who work with wood and other building materials to construct, install, and repair structures, frameworks, and fixtures.

Electrician: Electricians are responsible for installing, maintaining, and repairing electrical systems in buildings and construction projects.

Plumber: Plumbers handle the installation and maintenance of plumbing systems, including water supply, drainage, and sewer systems.

Support Roles

Estimator: Estimators assess project requirements and prepare accurate cost estimates for materials, labor, and other expenses. Their work is essential for bidding on projects and ensuring profitability.

Procurement Manager: The procurement manager oversees the purchasing and sourcing of materials and equipment for construction projects. They negotiate with suppliers, manage vendor relationships, and ensure timely delivery of materials.

Human Resources Manager: The HR manager is responsible for recruiting, hiring, and training employees. They also manage employee benefits, performance evaluations, and workplace policies.

Health and Safety Officer: Health and safety officers ensure that construction sites comply with safety regulations and implement measures to prevent accidents and injuries.

Quality Control Inspector: Quality control inspectors monitor construction projects to ensure that workmanship and materials meet quality standards and project specifications.

Administrative Assistant: Administrative assistants provide support to the management team and project managers by handling administrative tasks, managing schedules, and coordinating communications.

Each role within a construction business is critical to the overall success of the company. Identifying and defining the key roles allows construction companies to build a well-rounded team that can efficiently manage projects, deliver high-quality work, and maintain a safe and productive work environment. By recruiting the right talent for each role, construction businesses can foster a collaborative and skilled workforce that contributes to their growth and reputation within the industry. Effective communication and clear delineation of responsibilities are key to ensuring that each team member understands their role and contributes to the company's shared goals and objectives.

5.2 Hiring Skilled and Reliable Employees

Hiring skilled and reliable employees is crucial for the success of any construction business. A highly competent and dedicated workforce not only ensures the timely and efficient completion of construction projects but also contributes to the company's reputation and client satisfaction. To attract and retain top talent, construction businesses should implement effective hiring strategies that focus on identifying individuals with the necessary skills, experience, and commitment to excellence.

One of the primary steps in hiring skilled and reliable employees is to develop a comprehensive job description for each position. Clearly outlining the responsibilities, qualifications, and expectations helps attract candidates who possess the required skills and experience. The job description should highlight the specific technical skills, certifications, and industry knowledge necessary for the role.

To source qualified candidates, construction businesses can utilize a variety of methods. These include online job boards, professional networking platforms, industry-specific websites, and local trade schools or technical colleges. By targeting these channels, companies can reach a broader pool of potential employees who have a specific interest or background in the construction industry.

When evaluating candidates, it's essential to conduct thorough interviews to assess their skills, experience, and fit within the company culture. Behavioral and situational questions can help determine a candidate's problem-solving abilities, communication skills, and ability to work effectively in a team. Additionally, requesting references and conducting background checks provide valuable insights into a candidate's work history and reliability.

Skill assessments and practical tests can be valuable tools for evaluating a candidate's technical abilities. For example, a carpenter candidate may be asked to demonstrate their proficiency in specific woodworking techniques or the ability to interpret construction blueprints. These assessments help validate the candidate's skills and ensure they can perform the required tasks to a high standard.

While technical skills are crucial, assessing a candidate's attitude, work ethic, and commitment to safety is equally important. Construction businesses should prioritize hiring individuals who demonstrate a strong work ethic, attention to detail, and a commitment to following safety protocols. Conducting behavioral assessments and asking situational questions related to safety and workplace practices can help gauge a candidate's mindset and dedication to maintaining a safe work environment.

During the hiring process, it's crucial to communicate the company's values, culture, and opportunities for growth and development. Emphasizing the company's commitment to employee training and career advancement can attract candidates seeking long-term opportunities. Construction businesses should highlight their commitment to safety, quality workmanship, and fostering a collaborative and supportive work environment.

To retain skilled and reliable employees, construction businesses should provide competitive compensation packages that include fair wages, benefits, and opportunities for advancement. Regular performance evaluations and feedback sessions help employees understand their strengths, identify areas for improvement, and set goals for professional development. Offering ongoing training and certification programs not only enhances employee skills but also demonstrates a commitment to their growth and career progression.

Establishing a positive work culture that values teamwork, respect, and open communication is crucial for retaining skilled employees. Encouraging employee engagement and providing opportunities for employees to contribute ideas and suggestions fosters a sense of ownership and loyalty. Recognizing and rewarding exceptional performance and providing a safe and supportive work environment further reinforce the commitment to employee satisfaction.

Construction businesses can also consider implementing mentorship programs, where experienced employees provide guidance and support to new hires. This helps facilitate knowledge transfer, accelerates skill development, and strengthens team cohesion.

Hiring skilled and reliable employees is a critical aspect of building a successful construction business. By developing comprehensive job descriptions, utilizing targeted recruitment channels, conducting thorough interviews, and assessing technical skills and attitude, companies can attract candidates who meet the requirements of each position. To retain skilled employees, construction businesses should provide competitive compensation packages, opportunities for growth and development, and a positive work culture. By investing in their employees' skills, career progression, and job satisfaction, construction businesses can build a talented and dedicated workforce that contributes to the company's success and reputation within the industry.

5.3 Employee Training and Development

Employee training and development play a crucial role in the success and growth of a construction business. A well-trained and skilled workforce not only improves the quality and efficiency of construction projects but also enhances employee morale, retention, and overall job satisfaction. By investing in training and development programs, construction companies can foster a culture of continuous learning, innovation, and adaptability, which is essential in a dynamic and ever-evolving industry.

One of the primary benefits of employee training is the improvement of technical skills. Construction projects require specialized knowledge and expertise in various trades, such as carpentry, electrical work, plumbing, and masonry. Providing employees with opportunities for training and upskilling ensures that they remain current with industry best practices and advancements in technology and construction methods. This, in turn, leads to higher-quality work and increased customer satisfaction.

Employee training also plays a significant role in workplace safety. The construction industry is inherently risky, with potential hazards ranging from falls and equipment accidents to exposure to hazardous materials. Proper safety training equips employees with the knowledge and skills to identify and mitigate risks, adhere to safety protocols, and respond effectively in emergency situations. By prioritizing safety training, construction companies can reduce the likelihood of workplace accidents and injuries, creating a safer and more productive work environment.

In addition to technical skills and safety training, employee development should also focus on soft skills, such as communication, leadership, and teamwork. Effective communication is vital in the construction industry, where project success often hinges on clear and timely communication between various team members, contractors, and clients. Leadership training prepares employees to take on supervisory roles and effectively manage teams, projects, and resources.

Teamwork training fosters collaboration, trust, and synergy among employees, leading to better project coordination and outcomes.

Employee training and development can take various forms, ranging from on-the-job training and workshops to seminars, webinars, and online courses. On-the-job training allows employees to learn by doing and provides real-world experience under the guidance of experienced colleagues. Workshops and seminars led by industry experts or trainers offer opportunities for in-depth learning and knowledge sharing on specific topics. Online courses provide flexibility for employees to access training materials at their convenience, making it easier to balance work and learning commitments.

Mentorship programs are also effective in employee development. Pairing less experienced employees with seasoned professionals allows for knowledge transfer, skill development, and career guidance. Mentorship programs foster a sense of camaraderie and support within the organization, contributing to higher employee engagement and retention.

To ensure the effectiveness of employee training and development programs, construction businesses should conduct regular assessments and evaluations. Feedback from employees, supervisors, and clients can provide valuable insights into the impact of training initiatives on job performance and project outcomes. By continuously evaluating and refining training programs, construction companies can optimize their investment in employee development and ensure that training aligns with organizational goals and industry requirements.

Employee training and development contribute significantly to employee retention and job satisfaction. Providing opportunities for professional growth and skill development signals to employees that the company values their contributions and is committed to their long-term success. Employees who feel supported and invested in by their employer are more likely to remain loyal and motivated to excel in their roles.

Employee training and development are essential components of building a skilled, safe, and engaged workforce in the construction industry. Investing in employee development improves employee retention, job satisfaction, and overall organizational performance. Through a commitment to continuous learning and professional growth, construction companies can adapt to industry changes, stay ahead of the competition, and achieve long-term success in the dynamic and challenging construction sector.

5.4 Fostering a Positive Work Culture and Team Dynamics

Fostering a positive work culture and team dynamics is essential for creating a collaborative and productive environment in a construction business. A positive work culture is characterized by open communication, mutual respect, shared values, and a focus on employee well-being. When employees feel supported, valued, and engaged, they are more likely to work together effectively, take pride in their work, and contribute to the company's success.

One of the foundational elements of a positive work culture is effective leadership. Strong and empathetic leadership sets the tone for the entire organization and influences employee behavior and attitudes. Leaders who lead by example, communicate transparently, and actively listen to employee concerns build trust and credibility. They motivate and inspire their teams, encouraging creativity, innovation, and continuous improvement.

Communication is a vital aspect of fostering a positive work culture and team dynamics. Transparent and open communication ensures that employees are well-informed about company goals, projects, and changes within the organization.

Regular team meetings, one-on-one discussions, and feedback sessions create opportunities for employees to share their ideas, voice concerns, and collaborate on solutions. Constructive feedback and recognition for a job well done reinforce the sense of teamwork and value each employee brings to the organization.

In a positive work culture, employees are encouraged to take ownership of their work and are empowered to make decisions within their areas of expertise. Autonomy and responsibility instill a sense of pride and accountability in employees, leading to higher job satisfaction and commitment to achieving organizational goals.

Recognizing and celebrating employee achievements, both big and small, is essential for building team morale and motivation. Acknowledging hard work and accomplishments publicly or through rewards and incentives demonstrates that the company values and appreciates its employees' efforts.

To foster a positive work culture and team dynamics, it is crucial to promote diversity and inclusion. Embracing diversity in the workforce and fostering an inclusive environment enhances creativity, problem-solving, and decision-making. It ensures that all employees feel welcomed and respected, regardless of their background or identity.

Investing in employee well-being is also key to maintaining a positive work culture. Providing a safe and healthy work environment, offering wellness programs, and encouraging work-life balance are essential for employee satisfaction and long-term retention. When employees feel cared for and supported, they are more likely to be engaged and productive.

Team-building activities and events are effective ways to strengthen team dynamics and foster a sense of camaraderie among employees. Organizing team-building workshops, retreats, or volunteer activities promotes collaboration, communication, and trust among team members. These activities create opportunities for employees to get to know each other on a personal level, leading to better teamwork and cooperation.

Leadership development programs are valuable for building strong team dynamics within a construction business. Providing leadership training and mentoring opportunities helps develop future leaders within the organization. When employees see a clear path for growth and advancement, they are more motivated to invest in their professional development and contribute to the team's success.

Conflict resolution is a crucial skill for maintaining positive team dynamics. Conflict is a natural part of any workplace, and how it is managed can significantly impact team cohesion and productivity. Encouraging open communication and providing resources for conflict resolution and mediation help employees address issues constructively and collaboratively.

CHAPTER 6: SAFETY AND INSURANCE

6.1 Prioritizing Safety in Construction Projects

Prioritizing safety in construction projects is of paramount importance to protect the well-being of workers and ensure the successful execution of the project. The construction industry is inherently hazardous, with potential risks ranging from falls and equipment accidents to exposure to hazardous materials. Therefore, implementing a comprehensive safety program and fostering a safety-focused culture are crucial in mitigating these risks and creating a safe work environment.

One of the primary ways to prioritize safety in construction projects is to develop and enforce robust safety policies and procedures. These policies should address potential hazards and risks specific to each construction site, as well as compliance with relevant safety regulations and standards. Regular safety inspections, conducted by qualified safety professionals, help identify potential hazards and ensure that safety protocols are being followed.

Safety training is another critical aspect of prioritizing safety in construction projects. All workers, from skilled tradespeople to supervisors and project managers, should receive comprehensive safety training before starting work on a construction site. This training should cover the proper use of personal protective equipment (PPE), safe work practices, emergency procedures, and hazard recognition. Regular refresher courses and ongoing safety education are also essential to reinforce safe behaviors and keep safety measures top of mind.

Empowering workers to actively participate in safety initiatives is vital in creating a safety-focused culture. Encouraging employees to report safety concerns, hazards, and near-miss incidents fosters a sense of ownership and responsibility for safety. This information allows management to address potential issues proactively and make improvements to the safety program.

Ensuring that all equipment and tools are properly maintained and in good working condition is crucial for safety in construction projects. Regular inspections and maintenance checks for equipment such as cranes, scaffolding, and power tools help prevent accidents and equipment failures on the construction site.

In addition to worker safety, construction projects should prioritize public safety. Erecting appropriate barriers, signage, and warning systems around construction sites helps protect pedestrians and nearby properties from potential hazards.

To further prioritize safety, construction businesses should conduct thorough risk assessments for each project. Identifying potential risks and hazards at the planning stage allows for the implementation of preventive measures and risk mitigation strategies. This proactive approach can significantly reduce the likelihood of accidents and injuries throughout the construction process.

Creating a safety committee or appointing a safety officer responsible for overseeing safety initiatives and addressing safety concerns can help ensure safety measures are consistently enforced on construction sites. This committee or officer can also be instrumental in conducting safety meetings, audits, and fostering a safety-conscious culture.

Recognizing and rewarding safe behavior and accomplishments in safety is an effective way to reinforce the importance of safety among workers. Incentive programs that acknowledge and celebrate safety milestones and practices can encourage employees to remain vigilant about safety.

Employing technology and innovations can also enhance safety in construction projects. Implementing wearable safety devices, such as smart helmets or vests, can track worker movements and monitor vital signs, providing real-time data to enhance safety measures and respond quickly to emergencies.

Safety should be integrated into every stage of a construction project, from planning and design to construction and completion. By conducting safety reviews and risk assessments during the planning phase, construction businesses can identify and address potential safety concerns early on, optimizing safety measures and reducing the likelihood of accidents during the construction process.

6.2 Understanding Different Types of Insurance Coverage

Understanding different types of insurance coverage is crucial for protecting your construction business and mitigating financial risks. The construction industry involves various potential risks and liabilities, including property damage, injuries, and legal claims. Having appropriate insurance coverage provides financial protection and peace of mind in the event of unforeseen circumstances. Here are some key types of insurance coverage that construction businesses should consider:

General Liability Insurance: General liability insurance is essential for construction businesses as it provides coverage for third-party bodily injury, property damage, and personal injury claims. This insurance protects against accidents that may occur on the construction site or as a result of construction activities. It covers legal costs, medical expenses, and potential settlements or judgments.

Workers' Compensation Insurance: Workers' compensation insurance is a legal requirement in many jurisdictions. It provides coverage for work-related injuries or illnesses that employees may suffer during the course of their employment. Workers' compensation insurance covers medical expenses, lost wages, and rehabilitation costs for injured employees. It also protects the construction business from potential lawsuits related to workplace injuries.

Contractor's All Risk (CAR) Insurance: CAR insurance is specifically designed for construction projects. It provides coverage for property damage and loss caused by accidents, theft, vandalism, fire, or natural disasters. CAR insurance typically covers construction materials, equipment, and machinery, as well as any third-party property damage that may occur during construction activities.

Professional Liability Insurance: Professional liability insurance, also known as errors and omissions insurance, is crucial for construction businesses that provide professional services, such as design or consulting. It protects against claims of professional negligence, errors, or omissions that may result in financial losses for clients. Professional liability insurance covers legal defense costs, settlements, or judgments arising from such claims.

Commercial Auto Insurance: If your construction business owns or operates vehicles for business purposes, commercial auto insurance is necessary. This insurance provides coverage for accidents, property damage, and injuries involving company-owned vehicles. It also covers theft, vandalism, and damage to the vehicles.

Builder's Risk Insurance: Builder's risk insurance is specific to construction projects and covers the property being built or renovated. It provides protection against damage or loss caused by perils such as fire, vandalism, theft, or severe weather events. Builder's risk insurance typically includes coverage for materials, equipment, and the structure itself during the construction phase.

Environmental Liability Insurance: Environmental liability insurance is important for construction businesses that may face potential environmental risks, such as pollution or contamination. This insurance provides coverage for cleanup costs, legal expenses, and damages associated with environmental claims.

Cyber Liability Insurance: In the digital age, cyber liability insurance is becoming increasingly important for construction businesses. This insurance protects against data breaches, cyberattacks, and the resulting financial losses. It covers costs associated with data recovery, legal fees, and notification and credit monitoring services for affected parties.

It is crucial to carefully review insurance policies, understand the coverage limits, deductibles, and exclusions associated with each type of insurance. Consulting with an insurance professional who specializes in construction insurance can help you identify the specific risks your business faces and ensure you have adequate coverage in place.

6.3 Risk Management Strategies for Construction Businesses

Contingency planning is a crucial risk management strategy that involves identifying potential risks and developing alternative courses of action to address them. This proactive approach allows construction businesses to be prepared for unexpected events that could impact project timelines or budgets. Having contingency plans in place helps minimize disruptions and ensures that projects can proceed smoothly even in the face of unforeseen challenges.

Effective project management is another key risk management strategy for construction businesses. This includes establishing clear project objectives, creating detailed project plans, and defining roles and responsibilities. Regular project monitoring and performance tracking allow construction businesses to identify potential risks or deviations from the plan early on, enabling timely corrective action.

Supplier and subcontractor management is critical for construction businesses, as the performance of external parties can directly impact project outcomes. Conducting due diligence and selecting reliable suppliers and subcontractors can reduce the risk of delays or quality issues. Having clear contracts and communication channels with these external parties is essential for ensuring smooth collaboration and project execution.

Adopting technology and construction software can significantly enhance risk management in construction projects. Project management tools, scheduling software, and building information modeling (BIM) technology streamline communication, increase transparency, and facilitate data-driven decision-making. Integrating technology into construction processes improves efficiency and minimizes the risk of errors or miscommunications.

Regular communication and collaboration among all project stakeholders are vital for effective risk management. This includes regular project meetings, progress updates, and open channels for feedback and concerns. Effective communication fosters a sense of teamwork and ensures that everyone is aligned with project objectives and risk mitigation strategies.

Environmental risks are also significant considerations in construction projects. Understanding and complying with environmental regulations, conducting environmental impact assessments, and implementing environmentally friendly practices are essential for managing environmental risks and promoting sustainable construction.

Legal and contractual risks can pose significant challenges to construction businesses. Engaging legal counsel to review contracts, negotiate terms, and provide guidance on risk management can help protect the company's interests and avoid costly legal disputes.

Finally, learning from past projects and experiences is crucial for continuous improvement in risk management. Post-project evaluations and reviews allow construction businesses to identify lessons learned and implement improvements in future projects. A culture of learning and adaptability strengthens risk management practices and contributes to the long-term success of the construction business.

6.4 Ensuring Compliance with Occupational Safety Regulations

A robust safety program is essential to enforce safety regulations and ensure compliance on construction sites. This program should outline specific safety procedures, guidelines, and responsibilities for all employees and contractors. It should also include a clear chain of command for reporting safety concerns and incidents, as well as procedures for addressing and documenting safety violations.

Training and education play a crucial role in promoting safety compliance. All employees, including new hires and subcontractors, should receive thorough safety training before starting work on the construction site. This training should cover hazard recognition, safe work practices, proper use of personal protective equipment (PPE), and emergency procedures. Regular refresher courses and ongoing safety education are essential to reinforce safe behaviors and keep safety measures top of mind.

Construction businesses should conduct regular safety inspections and audits to identify potential hazards and ensure compliance with safety regulations. These inspections should be conducted by qualified safety professionals or trained personnel who can assess the construction site for compliance with safety standards and best practices. Any safety issues or violations should be promptly addressed and corrected.

Promoting a safety culture within the organization is vital for maintaining compliance with safety regulations. This involves fostering an environment where safety is a top priority and where employees feel empowered to speak up about safety concerns without fear of reprisal. Recognizing and rewarding safe behaviors and achievements can also reinforce the importance of safety among employees.

Communication is key to ensuring compliance with safety regulations. Regular safety meetings and toolbox talks provide opportunities for open discussions about safety concerns and updates on safety measures. Clear and consistent communication between management, supervisors, and workers helps ensure that everyone is aware of safety requirements and expectations.

Construction businesses should also keep abreast of changes in safety regulations and update their safety program accordingly. Staying informed about industry best practices and new safety technologies can further enhance safety compliance and overall safety performance.

Implementing safety incentives and recognition programs can motivate employees to prioritize safety. Rewarding employees for demonstrating exemplary safety practices or reporting safety hazards encourages a proactive safety culture and reinforces the value of safety compliance.

Partnering with safety consultants or hiring a dedicated safety officer can provide construction businesses with specialized expertise and guidance in safety compliance. These professionals can assist with safety program development, conduct safety audits, and ensure that the business stays up-to-date with the latest safety regulations.

Ensuring compliance with occupational safety regulations is a fundamental responsibility for construction businesses. Establishing a comprehensive safety program, providing thorough training, conducting regular inspections, and promoting a safety-conscious culture are key steps in achieving compliance. Clear communication, staying informed about regulatory changes, and partnering with safety experts are additional strategies to enhance safety compliance in the construction industry. By prioritizing safety and compliance, construction businesses can protect their workforce, mitigate risks, and maintain a positive reputation within the industry.

CHAPTER 7:
EQUIPMENT AND
SUPPLIES

7.1 Choosing the Right Equipment for Different Projects

C hoosing the right equipment for different construction projects is essential to ensure efficiency, productivity, and successful project outcomes. The construction industry relies heavily on various types of equipment to perform tasks ranging from excavation and material handling to concrete pouring and finishing. Selecting the appropriate equipment for each project requires careful consideration of factors such as project scope, site conditions, budget constraints, and timeline. Here are some key considerations and guidelines for choosing the right equipment for different construction projects:

Project Scope and Requirements: Understanding the specific requirements of the construction project is the first step in selecting the right equipment. The type of construction, size of the project, and scope of work will determine the equipment needs. For example, a large-scale commercial building project may require heavy machinery like excavators and cranes, while a smaller residential project may rely on smaller equipment like skid steers and mini-excavators.

Site Conditions: Site conditions play a crucial role in equipment selection. Factors such as terrain, soil type, weather conditions, and available space can impact the performance and suitability of equipment. For challenging terrains or limited access areas, compact and versatile equipment may be more suitable, while stable ground conditions may allow for larger and more specialized machinery.

Equipment Capability and Versatility: Assessing the capability and versatility of equipment is essential for ensuring that it can handle various tasks required in the project. Some equipment may have specific features or attachments that make them more versatile and adaptable to different construction tasks. For instance, a backhoe loader with different attachments can perform excavation, loading, and backfilling tasks, making it a valuable multi-purpose machine.

Equipment Availability and Rental Options: Consider the availability of the equipment in the local market and the possibility of rental options. Renting equipment can be a cost-effective solution for projects that require specialized machinery or have a limited duration. Evaluating the costs of ownership versus rental can help make informed decisions.

Safety and Operator Training: Prioritize equipment that meets safety standards and has modern safety features. Ensuring that operators are adequately trained to operate the equipment safely is crucial for preventing accidents and maximizing productivity. Investing in operator training can improve efficiency and reduce the risk of equipment-related incidents.

Maintenance and Support: Consider the maintenance requirements and availability of support for the selected equipment. Regular maintenance and timely repairs are essential for keeping equipment in optimal working condition and minimizing downtime. Choose equipment from reputable manufacturers with a strong support network for spare parts and technical assistance.

Budget Constraints: Construction projects often have budget constraints, and equipment costs can significantly impact the overall budget. Balancing the need for quality and reliable equipment with budget limitations is essential. Seeking cost-effective options that meet project requirements can help manage expenses without compromising on quality.

Environmental Considerations: In modern construction practices, sustainability and environmental considerations are crucial. Choosing equipment that is energy-efficient, meets emission standards, and has low noise levels can contribute to a more environmentally friendly construction site.

Choosing the right equipment for different construction projects is a strategic decision that involves considering project scope, site conditions, equipment capability, budget constraints, and safety considerations. A thorough understanding of project requirements, along with the availability of equipment and rental options, helps make informed decisions. Prioritizing safety, operator training, and equipment maintenance ensures smooth project execution and minimizes downtime. Taking environmental factors into account contributes to sustainable construction practices. Through careful evaluation these factors and making informed choices, construction businesses can enhance productivity, reduce costs, and achieve successful project outcomes.

7.2 Evaluating Equipment Suppliers and Rental Options

When evaluating equipment suppliers and rental options, it is crucial to consider the availability and condition of the equipment. The suppliers should have a sufficient inventory of the required equipment to meet project demands. Additionally, inspect the equipment to ensure it is in good working condition and meets safety standards. Well-maintained equipment is less likely to cause delays or breakdowns during the project.

Cost is another important consideration. Obtain quotes from multiple suppliers and rental companies to compare prices. However, it is essential to balance cost with the quality and reliability of the equipment. Opting for the cheapest option may result in subpar equipment that can lead to delays and additional costs in the long run. Consider the overall value provided by the supplier, including equipment quality, reliability, customer service, and support.

Flexibility in rental options is valuable, especially for projects with fluctuating equipment needs. Look for suppliers and rental companies that offer flexible rental terms, such as daily, weekly, or monthly rates, to accommodate project timelines and budget constraints. Additionally, inquire about the process for extending rental periods or exchanging equipment if project requirements change.

Prompt and responsive customer service is crucial in the construction industry, where time is of the essence. Evaluate the supplier's ability to provide timely delivery and pick-up of equipment. Clear communication channels and a dedicated point of contact can streamline coordination and address any issues or concerns promptly.

Consider the proximity of equipment suppliers and rental companies to the project site. Choosing suppliers located near the construction site can reduce transportation costs and minimize project delays due to equipment availability. It also allows for easier communication and coordination throughout the project duration.

Equipment suppliers and rental companies should have proper insurance coverage, including liability insurance. This coverage protects the construction business from any damages or liabilities that may arise from the use of rented equipment. Request proof of insurance from the suppliers and rental companies before finalizing any agreements.

Take into account the supplier's knowledge and expertise in the construction industry. Experienced suppliers can provide valuable guidance on equipment selection, usage, and maintenance. They can also offer recommendations on the most suitable equipment for specific project requirements.

Seek recommendations and references from other construction professionals or industry associations. Their insights and experiences can provide valuable information about the reliability and performance of equipment suppliers and rental companies.

Finally, review the terms and conditions of the rental agreement or purchase contract carefully. Pay attention to factors such as deposit requirements, equipment return conditions, and liability clauses. Ensure that the terms align with the project needs and protect the interests of the construction business.

7.3 Managing and Maintaining Construction Equipment

Regular equipment inspections are critical for identifying potential issues and ensuring that equipment is safe and in good working condition. Establish a schedule for routine inspections, and conduct pre-use checks before starting each workday. Inspections should include checking for signs of wear, leaks, loose or damaged parts, and ensuring that safety features are functioning correctly. Any identified issues should be addressed promptly through repairs or maintenance to prevent more significant problems.

Proper maintenance is key to extending the lifespan of construction equipment and reducing the likelihood of breakdowns. Develop a preventive maintenance program that includes regular servicing, lubrication, and replacement of consumable parts. Follow the manufacturer's maintenance guidelines and schedule to ensure that equipment operates at its optimal performance.

Documentation is essential for managing and maintaining construction equipment. Keep detailed records of equipment inspections, maintenance activities, and repairs. This documentation helps track the equipment's history, aids in troubleshooting, and provides valuable information when deciding on the timing of equipment replacement.

Implement a system to track equipment usage and productivity. Monitoring equipment usage helps identify high-demand equipment and determine if additional units are needed to meet project requirements. It also helps identify underutilized equipment that may be better suited for other projects or available for rent to generate additional revenue.

Training employees on equipment operation and maintenance is vital for safe and efficient equipment usage. Ensure that operators receive proper training on each piece of equipment they operate. Additionally, provide training on safety protocols, including proper use of personal protective equipment (PPE) and adherence to safety guidelines during equipment operation.

Create a dedicated storage and parking area for equipment when not in use. Proper storage protects equipment from weather elements and reduces the risk of theft or vandalism. Encourage operators to clean equipment after use to prevent dirt and debris from accumulating, which can lead to premature wear and damage.

Establish clear guidelines for reporting equipment issues or malfunctions. Encourage employees to report any problems they encounter with equipment promptly. Prompt reporting allows for timely repairs and prevents minor issues from developing into more significant and costly problems.

Consider the option of telematics technology for construction equipment. Telematics systems allow construction businesses to monitor equipment remotely, providing real-time data on equipment usage, location, and performance. This technology can help optimize equipment utilization, track maintenance needs, and improve overall fleet management.

Regularly review equipment performance and cost data to assess the efficiency of each piece of equipment. This data can help identify potential cost-saving opportunities and inform decisions about equipment replacement or upgrades.

7.4 Sourcing Quality Construction Materials and Supplies

Sourcing quality construction materials and supplies is a crucial aspect of any construction project. The use of high-quality materials directly impacts the durability, safety, and overall performance of the structures being built. Construction businesses must adopt a systematic approach to identify reliable suppliers, assess material quality, and ensure timely delivery. Here are some key considerations and strategies for effectively sourcing quality construction materials and supplies:

Choosing the right suppliers is the first step in sourcing quality construction materials. Conduct thorough research to identify reputable suppliers with a proven track record of delivering high-quality materials. Consider factors such as the supplier's experience in the industry, customer reviews, and references from other construction professionals. Building long-term relationships with reliable suppliers can lead to better pricing and timely deliveries.

Clearly define the material specifications required for the project. This includes detailing the quality, quantity, size, and other specific characteristics of each material. Having precise material specifications ensures that suppliers understand the project's needs and can provide suitable materials that meet industry standards.

Request product testing data and certifications from suppliers to validate the quality and compliance of the materials. Look for materials that meet relevant industry standards and have been tested for performance and safety. Ensuring that the materials are certified by recognized authorities provides confidence in their suitability for the project.

Familiarize yourself with local building codes, regulations, and material standards that apply to your construction projects. Compliance with these requirements is essential to ensure the safety and legality of the structures being constructed. Choose materials that meet or exceed these standards.

Request samples or create mock-ups of construction materials to assess their quality and suitability for the project. This allows for hands-on evaluation of the materials' appearance, texture, and durability. It also provides an opportunity to visualize how the materials will look in the final construction.

Purchasing materials in bulk quantities can lead to cost savings. Negotiate pricing with suppliers to get competitive rates, especially for large orders. However, prioritize quality over price, as using subpar materials may lead to more significant costs in the long run due to repairs and replacements.

Ensure that suppliers can meet the project's timeline by providing timely deliveries of construction materials. Delays in material delivery can lead to project setbacks and increased costs. Communicate project deadlines clearly to suppliers and establish a schedule for material deliveries.

Consider the environmental impact of construction materials and opt for sustainable and eco-friendly options when possible. Green building materials promote sustainability, reduce waste, and enhance the project's overall environmental performance.

Cultivate strong relationships with suppliers to promote effective communication and collaboration. Maintaining good relationships with suppliers can lead to better service, preferential treatment, and access to new and innovative materials.

Implement a quality control process to inspect incoming materials and verify their adherence to specifications. Conduct regular checks during construction to ensure that materials are being used correctly and according to industry best practices.

CHAPTER 8: BIDDING AND ESTIMATION

8.1 The Importance of Accurate Project Estimation

Accurate project estimation is a cornerstone of successful construction management. It involves predicting the resources, time, and costs required to complete a construction project. Whether it's a small renovation or a large-scale infrastructure development, accurate estimation is essential for project planning, budgeting, and resource allocation. The significance of accurate project estimation can be seen in various aspects of construction management:

Proper resource allocation is crucial for timely project completion. Accurate estimation helps identify the quantity and types of materials needed and ensures that the right amount of resources, such as construction equipment and labor, are available when required. This prevents delays and ensures smooth project progress.

Accurate project estimation is essential for determining project timelines. By assessing the scope of work and resource availability, project managers can establish realistic schedules for each phase of construction. Timely completion is vital in the construction industry, as delays can lead to additional costs, contractual disputes, and reputational damage.

Estimating construction projects accurately enables construction businesses to create competitive and compelling project proposals. Clients are more likely to choose a construction company that presents a well-structured and transparent estimate, giving them confidence in the project's feasibility and professionalism.

Effective risk management is facilitated by accurate project estimation. Identifying potential risks and uncertainties during the estimation phase allows project managers to devise mitigation strategies and allocate contingency funds. This proactive approach minimizes the impact of unforeseen challenges and keeps the project on track.

Clients and stakeholders demand accurate and transparent project estimates. Misleading or inaccurate estimates can erode trust and damage relationships, leading to disputes and legal issues. Transparent and accurate estimation builds credibility and fosters positive relationships with clients and other stakeholders.

Accurate project estimation sets the foundation for efficient cost control during construction. By comparing actual costs with estimated costs, project managers can identify cost discrepancies and take corrective measures. This cost control approach helps prevent budget overruns and ensures financial accountability.

Accurate project estimation supports informed decision-making throughout the construction process. As the project progresses, project managers can refer back to the initial estimates to evaluate performance, identify areas for improvement, and make data-driven decisions.

8.2 Understanding the Bidding Process

Understanding the bidding process is fundamental for construction businesses that seek to secure contracts for various projects. The bidding process is a competitive procedure in which contractors or construction firms submit proposals to compete for a construction project. It involves careful preparation, thorough evaluation, and adherence to specific requirements set by the project owner or client. A successful bid not only secures the project but also lays the groundwork for a profitable and mutually beneficial partnership. Here's a comprehensive overview of the key aspects of the bidding process:

1. Pre-Bidding Preparation

Before initiating the bidding process, construction businesses must conduct extensive pre-bidding preparation. This includes researching upcoming projects, identifying suitable opportunities, and understanding the client's requirements and expectations. As part of this process, businesses may attend pre-bid meetings to gain insights into the project scope, timeline, budget, and any specific criteria that must be met.

2. Reviewing Bid Documents

Bid documents, also known as request for proposal (RFP) or invitation to bid (ITB), contain detailed project information, specifications, and submission requirements. Construction businesses must thoroughly review these documents to understand the project's scope of work, technical specifications, contract terms, and evaluation criteria. Any questions or clarifications regarding the bid documents should be sought through formal channels.

3. Cost Estimation

One of the most critical aspects of the bidding process is accurately estimating the project's cost. Construction businesses must carefully analyze the project requirements, material costs, labor expenses, equipment needs, overheads, and potential risks. Creating a comprehensive and competitive bid requires a thorough understanding of cost estimation methodologies and industry pricing norms.

4. Developing the Bid Proposal

The bid proposal is a formal document that outlines the

construction business's approach to completing the project, its qualifications, and its pricing. It typically includes an executive summary, project overview, proposed timeline, cost breakdown, team qualifications, and any additional information required by the client. The bid proposal should be well-structured, clear, and persuasive to distinguish the construction business from its competitors.

5. Compliance and Legal Requirements
Compliance with legal requirements is essential in the bidding process. Construction businesses must ensure that their bids adhere to all applicable laws, regulations, and industry standards. Additionally, submitting the bid within the specified deadline and meeting all submission requirements is crucial to avoid disqualification.

6. Submitting the Bid
Once the bid proposal is complete, the construction business submits it according to the client's specified process. This may involve delivering physical copies of the proposal or submitting it electronically through an online bidding platform. Timely submission is critical, as late bids are usually rejected.

7. Evaluation and Selection
After the bid submission deadline, the client evaluates the proposals based on predefined criteria. The evaluation process may include technical evaluations, cost comparisons, and assessments of the construction business's qualifications and past performance. The client then selects the most suitable bid that best meets the project's requirements and provides value for money.

8. Post-Bid Follow-Up
After the bidding process, construction businesses often engage in post-bid follow-up with the client. This may involve seeking feedback on the proposal, discussing any clarifications,

or negotiating contract terms. Although the bidding process is competitive, maintaining professionalism and positive communication can lead to future opportunities with the client.

Understanding the bidding process is essential for construction businesses seeking to win contracts and secure profitable projects. Successful bidding requires thorough preparation, accurate cost estimation, compliance with legal requirements, and the development of compelling bid proposals. By carefully navigating the bidding process, construction businesses can position themselves competitively in the market, establish strong client relationships, and seize valuable opportunities for growth and success.

8.3 Factors to Consider in Preparing Bids for Construction Projects

Preparing bids for construction projects is a critical process that requires careful consideration and attention to detail. Bidding is the initial step in securing construction contracts, and successful bids can lead to profitable and rewarding projects for construction businesses. To increase the likelihood of winning bids, construction companies must thoroughly assess various factors during the bid preparation stage. Here are key factors to consider when preparing bids for construction projects:

1. Project Scope and Requirements

Understanding the project scope and requirements is essential before initiating the bidding process. Review the project documents, including the request for proposal (RFP) or invitation to bid (ITB), to gain insights into the project's objectives, technical specifications, and deliverables. Identifying the project's unique challenges and requirements is crucial for tailoring the bid proposal accordingly.

2. Cost Estimation and Budgeting

Accurate cost estimation is a fundamental aspect of bid

preparation. Construction businesses must analyze the project's scope and break it down into various components, such as labor, materials, equipment, subcontractors, and overhead costs. Utilize historical cost data, industry benchmarks, and expert knowledge to create a detailed and realistic budget for the project.

3. Timeline and Schedule

Consider the project timeline and schedule when preparing bids. Evaluate the feasibility of meeting the project's deadlines based on the available resources, workforce, and construction methodologies. Presenting a clear and achievable project timeline in the bid proposal demonstrates reliability and commitment to timely project completion.

4. Team and Expertise

Highlight the qualifications and expertise of the construction team in the bid proposal. Emphasize the relevant experience of key personnel, such as project managers, engineers, and supervisors. Demonstrating the team's capabilities and successful completion of similar projects can instill confidence in the client about the construction company's ability to deliver the project.

5. Risk Assessment and Mitigation

Identify potential risks associated with the project and develop strategies for risk mitigation. Addressing risks proactively in the bid proposal shows the construction business's understanding of the project's complexities and its commitment to delivering a successful outcome.

6. Compliance with Regulations and Standards

Ensure that the bid proposal complies with all applicable regulations, building codes, and industry standards. Non-compliance can lead to disqualification or legal issues. Demonstrating a commitment to adherence to regulations showcases the construction company's dedication to safety, quality, and ethical practices.

7. Subcontractors and Suppliers

Evaluate the need for subcontractors and suppliers to support the project. If subcontractors are required, select reputable and reliable partners with a track record of delivering quality work. Include details about subcontractors and suppliers in the bid proposal to demonstrate a well-rounded and capable team.

8. Value Engineering and Innovation

Consider value engineering and innovative solutions when preparing bids. Propose cost-effective and efficient alternatives without compromising quality. Clients appreciate construction companies that can offer innovative approaches that optimize project outcomes.

9. Clear and Concise Proposal

Present the bid proposal in a clear and concise manner. Avoid unnecessary jargon and provide all relevant information in a structured format. A well-organized and professional bid proposal enhances the construction company's image and increases its chances of being selected.

10. Competitive Pricing Strategy

Develop a competitive pricing strategy based on the project's requirements and the company's cost estimation. Be mindful of balancing competitive pricing with the need for profitability and quality. Offering value for money while staying competitive can set the bid apart from others.

8.4 Strategies for Competitive and Profitable Bidding

Competitive and profitable bidding is a crucial aspect of the construction industry. It involves submitting compelling proposals that not only secure projects but also ensure that the construction business remains financially viable and successful. To achieve competitive and profitable bidding, construction companies must adopt effective strategies that set them apart from competitors while maintaining a focus on profitability.

Thorough market research is the foundation of successful bidding. Construction businesses must closely monitor industry trends, upcoming projects, and the competitive landscape. Understanding the market dynamics allows companies to identify potential opportunities, assess client needs, and tailor their bids accordingly. Targeted project selection is also essential. Instead of bidding on every available project, construction companies should focus on projects that align with their expertise, resources, and capabilities. Targeted project selection increases the chances of winning contracts and ensures that the company can deliver high-quality results.

Building strong relationships with clients, subcontractors, and suppliers is essential for competitive bidding. Positive relationships instill confidence in clients and enhance the construction company's reputation. Repeat business and word-of-mouth referrals are often the result of fostering excellent relationships. Engaging with clients early in the project development process allows construction companies to gain insights into the project scope and requirements. Participating in pre-bidding meetings and discussions helps companies understand the client's expectations and build rapport.

Offering value engineering and innovative solutions can set a construction bid apart from competitors. Proposing cost-saving measures, improved construction methodologies, or sustainable practices demonstrates the company's commitment to efficiency and client satisfaction.

Accurate cost estimation is a fundamental aspect of competitive and profitable bidding. Construction businesses must analyze the project's scope and break it down into various components, such as labor, materials, equipment, subcontractors, and overhead costs. Utilize historical cost data, industry benchmarks, and expert knowledge to create a detailed and realistic budget for the project.

Maintaining a competitive pricing strategy is essential, but it must be balanced with the need for profitability. Low-balling bids to win projects may lead to financial strain and compromised quality. Construction companies must carefully assess their costs and set competitive prices that also allow for a reasonable profit margin.

Incorporating technological advancements and digital tools can streamline the bidding process and enhance efficiency. Utilize construction management software to organize bid documents, track project costs, and communicate with team members. Technology can also help with cost estimation, scheduling, and project coordination, contributing to better bids and project execution.

Continual improvement and learning from past bids are essential for refining bidding strategies. Conduct post-bid analyses to assess the strengths and weaknesses of each bid, gather feedback from clients, and identify areas for improvement. Incorporate these insights into future bids to enhance competitiveness and increase the likelihood of winning projects.

Collaborating with industry partners can strengthen a construction bid. Partnering with other construction companies or forming joint ventures can combine resources, expertise, and capabilities, enabling businesses to take on larger and more complex projects.

CHAPTER 9:
MARKETING AND
NETWORKING

9.1 Developing a Marketing Strategy
for Your Construction Business

Developing a marketing strategy for your construction business is essential to attract clients, increase brand visibility, and stay competitive in the industry. To create an effective marketing plan, you need to start by defining your goals and understanding your target audience. Knowing what you want to achieve through marketing and who your ideal clients are will guide your strategy.

Once you have a clear understanding of your goals and target audience, focus on building a strong brand identity. Your brand represents your company's values, mission, and unique selling points. Create a memorable logo, choose a cohesive color scheme, and develop a brand voice that aligns with your business's personality. Consistency in branding across all marketing channels will help establish brand recognition and credibility.

Having a professional website is crucial in today's digital world. Your website serves as the online face of your construction business. Ensure it is visually appealing, easy to navigate, and provides essential information about your services, past projects, and team. Include clear contact information and calls-to-action to encourage visitors to get in touch.

Content marketing is a powerful tool for construction businesses. Create valuable and informative content that addresses common questions and challenges in the construction industry. Blog posts, articles, videos, and infographics can showcase your expertise and help build trust with your audience. Share your content on your website, social media platforms, and industry-related forums.

Utilize social media to connect with potential clients and industry peers. Identify the social media platforms where your target audience is most active and engage with them regularly. Share updates about your projects, industry news, and informative content. Social media provides an opportunity to humanize your brand and build relationships with your audience.

Paid advertising can complement your organic marketing efforts. Consider using online advertising platforms like Google Ads or social media ads to reach a broader audience and generate leads. Target specific demographics and interests to ensure your ads are seen by the right people.

Networking and forming industry partnerships can also play a significant role in your marketing strategy. Attend industry events, join construction associations, and participate in networking opportunities to expand your connections. Collaborating with other professionals can lead to new opportunities and referrals.

Customer testimonials and case studies are powerful social proof for your construction business. Displaying positive feedback from satisfied clients and showcasing successful projects will build trust with potential clients and demonstrate your expertise.

Regularly measure and analyze the results of your marketing efforts. Use website analytics and social media insights to track your performance. Analyzing these metrics will help you understand what strategies are working and where adjustments are needed to improve your marketing approach.

9.2 Creating a Strong Online Presence and Website

Creating a strong online presence and a highly effective website is essential for the success of your construction business in today's digital age. In this competitive market, having a robust online presence will help you reach a broader audience, showcase your expertise, and attract potential clients. Here are some key steps to achieve a strong online presence and build an impactful website for your construction business:

Start by defining your construction company's brand identity. Understand your core values, mission, and unique selling propositions. Identify your target audience and research their needs and preferences. A well-defined brand identity will serve as the foundation for all your online efforts.

Next, focus on developing a professional website. Your website is the virtual storefront of your construction business, and it should create a positive first impression for potential clients. Ensure the website is visually appealing, user-friendly, and mobile-responsive. Highlight your services, portfolio of past projects, and provide clear contact information.

Showcasing your projects is crucial in demonstrating your capabilities and expertise. Include high-quality images, detailed project descriptions, and any awards or recognition received. This will help potential clients visualize the quality of your work and build trust in your construction business.

Provide valuable content on your website to position yourself as an industry authority. Consider starting a blog where you can share construction tips, industry trends, and insights. By offering valuable content, you can attract organic traffic to your website and establish your construction business as a trusted resource.

Utilize search engine optimization (SEO) techniques to improve your website's visibility on search engines. Research relevant keywords and integrate them into your website's content. Focus on creating informative and relevant content that aligns with what your target audience is searching for.

Embrace the power of social media to expand your online reach. Identify the platforms where your target audience is most active and engage with them regularly. Share updates about your projects, industry news, and behind-the-scenes glimpses of your construction work. Social media offers a great platform to interact with potential clients and showcase your expertise.

Leverage online reviews and testimonials from satisfied clients. Positive reviews and testimonials can significantly impact potential clients' decision-making process. Encourage your happy clients to leave reviews on platforms like Google My Business and social media.

Invest in professional photography and videography to visually represent your construction projects. High-quality visuals can make a powerful impression on potential clients and showcase your attention to detail and craftsmanship.

Consider incorporating video content on your website and social media channels. Video is a highly engaging format and can be used to showcase your projects, introduce your team, and share construction tips and insights.

Engage with your audience through email marketing. Build an email list of potential clients, industry contacts, and past clients. Send regular newsletters and updates to keep them informed about your latest projects and industry news.

Monitor and analyze the performance of your website and online presence. Use website analytics tools to track traffic, user behavior, and conversion rates. Analyze social media metrics to understand which content resonates with your audience the most. Use these insights to refine and improve your online strategy over time.

9.3 Utilizing Social Media and Content Marketing

Social media and content marketing play a vital role in the success of your construction business. With the increasing influence of digital platforms, utilizing social media and content marketing strategies can help you reach a wider audience, establish your brand, and drive meaningful engagement. Here's how you can effectively leverage social media and content marketing for your construction business:

Social media platforms provide an excellent avenue to connect with your target audience and showcase your construction expertise. Create compelling profiles on platforms such as LinkedIn, Facebook, Instagram, or Twitter, tailored to your business goals and target audience.

Regularly share engaging and informative content related to the construction industry. This can include project updates, behind-the-scenes glimpses of your work, tips and tricks, and industry insights. Use visually appealing images and videos to captivate your audience and showcase the quality of your workmanship.

Engage with your followers by responding to comments, answering questions, and participating in relevant discussions. This interaction helps build relationships and establishes your business as a trusted authority in the construction field.

Content marketing is a powerful strategy to showcase your expertise, educate your audience, and drive traffic to your website. Start a blog on your website and create informative articles, guides, or case studies that address common construction challenges, offer valuable tips, or provide in-depth project analyses. This positions your business as a knowledgeable resource and helps build trust with potential clients.

Optimize your content for search engines by incorporating relevant keywords and using SEO best practices. This helps improve your website's visibility in search engine results, driving organic traffic to your site.

Consider creating video content, such as project walkthroughs, construction tutorials, or client testimonials. Videos are highly engaging and can showcase your skills and capabilities effectively. Share these videos on your website and social media platforms to increase brand visibility and engage with your audience.

Collaborate with industry influencers, trade organizations, or complementary businesses in the construction field. This allows you to tap into their audience base, expand your reach, and establish yourself as a credible player in the industry.

Use social media advertising to reach a broader audience and target specific demographics or locations relevant to your construction business. Platforms like Facebook Ads or LinkedIn Ads offer robust targeting options to maximize the impact of your advertising budget.

Track and analyze the performance of your social media and content marketing efforts. Use analytics tools to measure engagement, website traffic, and conversions. This data provides valuable insights into what content resonates with your audience and helps you refine your strategies for better results.

Remember to stay consistent with your social media presence and content creation. Establish a content calendar to plan and schedule your posts in advance. Regularly monitor and engage with your audience to maintain an active and responsive online presence.

9.4 Building Relationships and Networking in the Construction Industry

Building relationships and networking within the construction industry is crucial for the growth and success of your business. Establishing strong connections with industry professionals, suppliers, clients, and other stakeholders can lead to valuable opportunities, referrals, and collaborations. Here are some effective strategies to build relationships and network within the construction industry:

Develop a strong presence in industry associations, trade shows, and conferences. Attend relevant events and actively participate in networking activities. Engage in conversations, exchange contact information, and follow up with potential contacts after the event.

Join industry-specific organizations and associations that align with your construction business. These platforms provide opportunities to connect with like-minded professionals, stay updated on industry trends, and gain access to valuable resources and networking events.

Utilize online platforms and social media channels to expand your network. Join construction-related groups and forums on platforms like LinkedIn or Facebook. Engage in discussions, offer insights, and connect with individuals who share similar interests or expertise.

Nurture relationships with suppliers and subcontractors. Maintain regular communication, seek recommendations for new projects, and collaborate on joint marketing initiatives. Strong partnerships with reliable suppliers can enhance your project delivery and reputation.

Establish strategic alliances with complementary businesses in the construction industry. Identify businesses that offer complementary services, such as architects, engineers, or interior designers. Collaborate on joint projects or refer clients to one another, creating a win-win situation.

Attend local business events, such as chamber of commerce meetings or industry-specific seminars. These events provide opportunities to meet professionals from various industries, exchange business cards, and explore potential partnerships.

Develop a referral program where you incentivize satisfied clients, industry contacts, and partners to refer your construction services. Offer incentives such as discounts, referral bonuses, or exclusive benefits to encourage referrals.

Stay in touch with your existing and past clients. Maintain regular communication to foster long-term relationships and encourage repeat business. Provide valuable information, share industry updates, and seek feedback to continuously improve your services.

Utilize digital platforms and online networking tools to expand your reach. Connect with industry professionals on LinkedIn, follow industry influencers on social media, and engage in discussions to establish your expertise and expand your network.

Seek opportunities to collaborate with local organizations or participate in community development projects. Involvement in community initiatives not only enhances your brand reputation but also opens doors to networking opportunities within the local community.

Attend educational workshops or seminars to enhance your skills and knowledge in the construction industry. These events often attract industry experts and thought leaders, providing excellent networking opportunities.

Follow up with your contacts regularly. Send personalized emails, make phone calls, or schedule meetings to maintain relationships and explore potential collaboration or business opportunities.

Remember, building relationships and networking in the construction industry is a long-term investment. It requires consistency, genuine interest in others, and a willingness to offer support and value. By actively engaging with industry professionals, suppliers, clients, and other stakeholders, you can establish a strong network that contributes to the growth and success of your construction business.

CHAPTER 10:
MANAGING PROJECTS
SUCCESSFULLY

10.1 Effective Project Management Techniques

Effective project management techniques are crucial for the successful execution of construction projects. The construction industry is complex and involves various stakeholders, tasks, and timelines. Thorough planning is the foundation of successful project management. Develop a detailed project plan that outlines the project scope, objectives, deliverables, milestones, and timelines. Identify potential risks and challenges and devise mitigation strategies.

Effective communication is key to keeping all stakeholders informed and aligned. Establish clear lines of communication among team members, clients, subcontractors, and suppliers. Regularly update everyone on project progress, changes, and decisions.

Team collaboration is essential for the smooth flow of construction projects. Foster a collaborative work environment where team members can openly share ideas, address challenges, and work together to achieve project goals. Encourage regular meetings and cross-functional cooperation.

Project managers must have excellent leadership skills to

guide the project team effectively. Provide clear direction, set expectations, and motivate team members to perform at their best. Address conflicts promptly and promote a positive work culture.

Utilize project management software and tools to streamline project tasks and improve efficiency. These tools can help with scheduling, resource allocation, budget tracking, and document management. They also facilitate real-time collaboration among team members.

Regularly monitor project progress and performance against the project plan. Identify any deviations and take corrective actions to keep the project on track. Review milestones, budgets, and timelines to ensure project objectives are met.

Risk management is critical in the construction industry, where unexpected challenges can arise. Anticipate potential risks and develop contingency plans to address them. Regularly assess and update risk management strategies throughout the project.

Quality control and assurance are essential to deliver a successful project. Establish quality standards and conduct regular inspections to ensure work meets the required specifications. Address any quality issues promptly to maintain project integrity.

Resource management involves efficiently allocating and managing resources, including labor, materials, and equipment. Optimize resource usage to avoid delays and cost overruns while maintaining quality standards.

Documentation is vital for recording project details, decisions, and changes. Keep accurate and comprehensive records of all project-related information to facilitate smooth communication and provide a reference for future projects.

Regularly communicate with clients and stakeholders to keep them informed about project progress, milestones, and any changes. Address client concerns promptly and maintain a high level of customer service throughout the project.

Continuous improvement is essential in project management. Conduct post-project evaluations to identify areas for improvement and gather feedback from stakeholders. Implement lessons learned in future projects to enhance project delivery.

10.2 Planning and Scheduling Construction Projects

Proper planning and scheduling are essential for successful construction projects. It involves considering various factors to ensure the project is executed efficiently and meets the desired outcomes. Here are some key factors to consider:

Firstly, define the project scope and objectives. Clearly outline the goals, deliverables, and client requirements. This sets the foundation for the entire project and ensures everyone is on the same page.

Next, conduct a thorough assessment of the project requirements, including resources, materials, and manpower. Determine the necessary permits, approvals, and any specific regulations or codes that need to be followed.

Once the project requirements are identified, develop a detailed project plan. This includes breaking down the work into tasks, estimating the duration of each task, and determining the order of tasks based on dependencies. Consider using project management software to assist in organizing and tracking project tasks.

Assign responsibilities to team members based on their expertise and availability. Clearly communicate roles and expectations to ensure everyone understands their responsibilities.

Develop a realistic project schedule that considers the availability of resources, potential risks, and external factors that may impact the timeline. Set milestones to track progress and ensure timely completion of critical project phases.

Regularly review and update the project schedule as needed. Monitor progress, identify any delays or bottlenecks, and take necessary corrective actions. Effective communication is crucial to keep all stakeholders informed of any changes to the schedule.

Consider conducting a risk assessment to identify potential risks and develop strategies to mitigate them. This helps in proactive risk management and minimizes the impact of unforeseen events on the project schedule.

Collaboration and coordination among team members and subcontractors are essential for a smooth workflow. Regular meetings and clear communication channels facilitate effective collaboration and help address any issues or conflicts that may arise.

Regularly monitor project costs to ensure they align with the allocated budget. Track expenses, review cost projections, and implement cost-control measures to avoid budget overruns.

Implement a document management system to organize project-related documentation, including permits, contracts, and change orders. This ensures easy access to critical information and facilitates efficient communication.

Maintain flexibility in the schedule to accommodate unexpected changes or unforeseen circumstances. It is important to have contingency plans and the ability to adjust the schedule if necessary.

Regularly communicate with clients and stakeholders to provide project updates, address concerns, and manage expectations. Transparency and open communication contribute to a successful project outcome.

10.3 Handling Changes and Unexpected Challenges

Handling changes and unexpected challenges in construction projects requires adaptability, effective communication, and a problem-solving mindset. Here are some strategies to navigate through such situations:

One key strategy is to anticipate and plan for changes. While it's impossible to predict every challenge, experienced project managers conduct risk assessments and include contingency plans in their initial project planning. This allows for a more agile response when unexpected challenges arise.

Maintaining open and transparent communication among all stakeholders is crucial. Regularly update clients, team members, subcontractors, and suppliers about any developments that may impact the project. Clear communication about the reasons for changes, their implications, and proposed solutions helps manage expectations and fosters collaboration.

Adopting a solutions-oriented mindset is vital. Rather than dwelling on the problem, focus on finding practical solutions. Engage the project team in brainstorming sessions to explore different approaches and select the most viable option. Encouraging innovative thinking can lead to creative solutions.

Prioritizing safety and quality is non-negotiable. When faced with unexpected challenges, never compromise on safety protocols or the quality of work. Ensure that changes do not compromise the well-being of workers or the public. Uphold high-quality standards and ensure that any modifications align with project requirements.

Flexibility in scheduling and resource allocation is essential. Changes in construction projects can impact the project schedule and resource distribution. Project managers should be ready to adjust timelines and reallocate resources as needed. Utilize project management tools to quickly reorganize tasks and resources in response to changes.

Collaboration with the project team is crucial. Engage team members in decision-making processes when handling changes or unexpected challenges. Their input and expertise can provide valuable insights and help identify the most effective solutions. Foster a culture of collaboration and encourage open communication among team members.

Maintain a proactive approach to problem-solving. Actively seek solutions and address challenges promptly. This includes identifying potential risks and taking preventive measures to minimize their impact. Regularly assess the project's progress and proactively address any issues that may arise.

Learn from challenges and changes encountered throughout the project. Conduct post-project evaluations to identify lessons learned and areas for improvement. Document best practices and update project management processes to enhance future project delivery.

10.4 Ensuring Quality Control and Delivering on Time

Ensuring quality control and delivering projects on time are two critical aspects of successful construction management. Meeting high-quality standards and adhering to project timelines are essential for client satisfaction, reputation building, and overall project success. Here are some key strategies to achieve quality control and timely project delivery in the construction industry:

Comprehensive planning and scheduling form the foundation of quality control and timely delivery. It is important to develop a detailed project plan that outlines the project scope, objectives, deliverables, and milestones. Breaking down the work into smaller tasks and allocating realistic timeframes for each activity is crucial. Consider potential risks and contingencies in the schedule to avoid delays.

Clearly defined quality standards are essential for maintaining consistent quality throughout the project. Establish clear quality criteria and communicate them to all project stakeholders. These standards should encompass materials, workmanship, safety protocols, and adherence to building codes and regulations. Regular inspections and assessments should be conducted to ensure compliance with these standards.

Building an engaged and skilled workforce is vital for quality control and timely delivery. Invest in employee training and development programs to enhance their skills and knowledge. Motivate and empower your team to take ownership of their tasks and produce high-quality work. Effective leadership and regular communication with the team can foster a culture of accountability and excellence.

Maintain strong relationships with reliable suppliers to ensure the timely delivery of quality materials. Establish clear channels of communication and coordinate closely with suppliers to monitor material availability and address any potential delays. Having backup suppliers or contingency plans in place can mitigate the risk of material shortages or delays.

Leverage technology and automation to streamline construction processes and improve efficiency. Implement construction management software that can help track project progress, manage resources, and provide real-time updates. This technology can enhance coordination and communication among team members, leading to better quality control and on-time project delivery.

Conduct regular quality inspections at different stages of the project to identify and rectify any quality issues promptly. Collaborate with quality control teams to assess compliance with specifications and standards. Addressing any concerns early on can prevent costly rework and delays. Regular communication with subcontractors and trade partners is essential to ensure they also adhere to quality standards and project timelines.

Establish effective communication channels among all project stakeholders to facilitate timely decision-making and issue resolution. Clear and open communication promotes collaboration and enables prompt identification and resolution of quality-related concerns. Regular progress meetings and reporting mechanisms can help track project milestones and address any potential deviations.

Continuous improvement is key to achieving and maintaining quality control. Regularly evaluate project performance and outcomes, and solicit feedback from clients and stakeholders. Use lessons learned to refine processes, enhance quality control measures, and optimize project schedules for future projects.

CHAPTER 11: CUSTOMER SATISFACTION AND REVIEWS

11.1 Prioritizing Customer Satisfaction in Construction Projects

Prioritizing customer satisfaction is paramount in the construction industry. Satisfied clients not only lead to repeat business but also contribute to positive word-of-mouth referrals and a strong reputation. Here are some key strategies to prioritize customer satisfaction in construction projects:

Understanding client needs is crucial in achieving customer satisfaction. Take the time to listen and understand their specific requirements, project goals, and desired outcomes. This enables you to tailor the project to their expectations and deliver results that meet or exceed their needs.

Maintaining effective communication throughout the project is essential. Keep clients informed of progress, milestones, and any changes that may arise. Regularly update them on project developments and promptly address any questions or concerns they may have. Transparent and open communication fosters trust and ensures that clients feel involved and valued.

Setting realistic expectations from the outset is important for customer satisfaction. Be honest and transparent about project timelines, potential challenges, and budget considerations. Avoid making promises that cannot be fulfilled. When clients have realistic expectations, they are more likely to be satisfied with the final outcome.

Delivering high-quality workmanship is crucial for customer satisfaction. Employ skilled and experienced professionals who take pride in their work. Use quality materials and adhere to industry standards and best practices. Regularly inspect the progress to ensure that the work meets the agreed-upon quality standards.

Attention to detail is a key aspect of customer satisfaction. Pay close attention to even the smallest details of the project. Clients appreciate a meticulous approach, as it demonstrates that their project is being handled with care and precision. Addressing minor details can significantly impact the overall satisfaction of the client.

Demonstrate flexibility and adaptability in dealing with unexpected challenges. Construction projects can encounter unforeseen circumstances, and how you handle them can make a significant difference. Be proactive in communicating changes and offering viable solutions. Clients value contractors who can navigate through challenges efficiently and maintain project progress.

Timely project delivery is crucial for customer satisfaction. Develop a realistic project schedule and adhere to it diligently. Prioritize timely completion without compromising on quality. Delays can lead to frustration and dissatisfaction, so effective project management is crucial.

Take accountability for any mistakes or shortcomings in the project and work promptly to resolve them. Be responsive to client concerns and address any issues that arise during the project. Demonstrating a commitment to customer satisfaction builds trust and loyalty.

After project completion, conduct a post-project follow-up to gather feedback from the client. This shows that you value their input and provides an opportunity to address any remaining concerns. Use this feedback to identify areas for improvement and implement changes in future projects.

Provide exceptional customer service by going the extra mile whenever possible. Consider small gestures such as providing periodic project updates, offering maintenance tips, or addressing any additional client requests beyond the initial scope. These actions demonstrate your commitment to customer satisfaction and create a positive and memorable experience for clients.

11.2 Handling Client Feedback and Complaints

Handling client feedback and complaints is a crucial aspect of providing exceptional customer service in the construction industry. While every effort is made to meet client expectations, it is inevitable that some clients may express concerns or dissatisfaction at times. Constructive feedback can be an opportunity for growth and improvement, and addressing complaints effectively is essential for maintaining positive client relationships. Here are some key strategies for handling client feedback and complaints:

When clients provide feedback or express complaints, active listening is paramount. Allow them to voice their concerns without interruption, and make an effort to understand their perspective fully. Acknowledge their feelings and reassure them that their feedback is valued.

Maintain a calm and professional demeanor when receiving client feedback or complaints. Avoid becoming defensive or confrontational. Instead, adopt a problem-solving approach to address the issue constructively.

Respond to client feedback and complaints promptly. Timely communication shows that you take their concerns seriously and are committed to finding a resolution.

If the client has experienced a negative experience, offer a sincere apology and express empathy for any inconvenience or frustration they may have experienced. Demonstrating empathy shows that you care about their feelings and are committed to making things right.

Thoroughly investigate the client's feedback or complaint to understand the root cause of the problem. Collaborate with your team to gather all relevant information and assess the situation objectively.

Once you have a clear understanding of the issue, communicate potential solutions to the client. Be transparent about what steps you will take to address their concerns and prevent similar issues in the future.

After implementing solutions, follow up with the client to ensure they are satisfied with the resolution. This follow-up demonstrates your commitment to resolving the issue and maintaining a positive client relationship.

Learn from client feedback to identify areas for improvement in your processes and services. Use feedback as an opportunity to refine your approach and enhance the overall client experience.

Establish a system for recording and analyzing client feedback and complaints. Use this data to identify patterns and trends, allowing you to proactively address potential issues in the future.

Train your team in handling client feedback and complaints effectively. Provide them with the necessary tools and skills to respond to clients with empathy and professionalism.

Recognize the value of client feedback and complaints as opportunities for continuous improvement. Embrace a culture of learning and growth within your construction company.

Seek feedback from satisfied clients as well. Positive feedback is an affirmation of your team's hard work and can serve as powerful testimonials for future clients.

Finally, maintain transparency throughout the process of addressing client feedback and complaints. Keep clients informed of progress and any challenges encountered in resolving their concerns.

11.3 The Impact of Positive Reviews on Your Construction Business

Positive reviews play a significant role in shaping the success and reputation of a construction business. In today's digital age, online reviews hold immense power, influencing potential clients' decisions and overall brand perception. Let's explore the impact of positive reviews on a construction business:

Positive reviews serve as powerful testimonials of a construction company's capabilities and credibility. When potential clients see positive feedback from satisfied customers, it instills confidence in the company's ability to deliver quality work and exceptional customer service. These reviews can sway potential clients in favor of choosing your construction business over competitors.

Word-of-mouth marketing has evolved into online reviews. Positive reviews are akin to personal recommendations, carrying the same weight in terms of trust and influence. They can amplify your construction business's reputation and attract a broader audience.

In the construction industry, where projects often require substantial investments, clients seek assurance before choosing a contractor. Positive reviews can help potential clients feel more secure about their decision, especially when they witness successful projects and happy customers.

Online review platforms like Google, Yelp, and specialized construction websites have become go-to sources for clients researching construction companies. A strong online presence with positive reviews can improve your company's visibility and credibility. It may even lead to higher search engine rankings, making it easier for potential clients to find your business.

Positive reviews contribute to brand loyalty and repeat business. When clients are pleased with their experience, they are more likely to return for future projects or recommend your services to others. These loyal customers become brand advocates, organically promoting your construction business through positive word-of-mouth.

Client feedback through reviews offers valuable insights for continuous improvement. Constructive feedback in reviews can highlight areas where your construction business excels and areas that may need attention. By analyzing reviews, you can identify trends, address any recurring issues, and refine your services to better meet client needs.

Construction projects often involve complex decision-making processes, and clients appreciate transparency and open communication. Positive reviews often mention strong communication and responsiveness, indicating that your company is attentive to clients' concerns and keeps them informed throughout the project.

Positive reviews not only impact potential clients but also attract talented professionals to your construction business. Job seekers often explore online reviews to gauge a company's reputation and work environment. A positive reputation can attract skilled workers who want to be part of a successful and reputable organization.

Construction is a relationship-driven industry, and positive reviews help build trust with both clients and partners. Subcontractors and suppliers are also influenced by a company's reputation, and positive reviews can lead to stronger working relationships and better collaboration on projects.

Responding to positive reviews is equally important. Acknowledge and thank clients for their kind words. This demonstrates appreciation for their feedback and reinforces your commitment to customer satisfaction. It also shows potential clients that you value and engage with your customers.

11.4 Building Long-Term Client Relationships and Referral Business

Building long-term client relationships and referral business is an essential aspect of any successful enterprise. Cultivating a strong client base that not only returns but also recommends your products or services to others can significantly impact the growth and sustainability of your business. This article will explore various strategies and best practices that can help businesses foster enduring client relationships and encourage referrals.

First and foremost, delivering exceptional service is paramount. Going the extra mile to meet and exceed client expectations is a surefire way to build loyalty. Whether it's through high-quality products, outstanding customer support, or prompt issue resolution, showing genuine care for clients will leave a positive and lasting impression.

Trust and credibility are the foundation of any long-term relationship. Honesty, transparency, and consistency in delivering promises are crucial elements to gain and maintain trust. Clients are more likely to stay with a brand they can rely on and believe in.

Personalization is key in making clients feel valued and

appreciated. Each client is unique, with specific needs and preferences. Tailoring products or services to address their individual requirements demonstrates that your business cares about their satisfaction.

Regular communication is vital to staying connected with clients beyond the initial transaction. Keeping them updated with relevant information, offering helpful content, and being available to address queries and concerns can foster a sense of engagement and loyalty.

Seeking feedback from clients is not only a sign of respect but also an opportunity for improvement. Actively encouraging clients to share their opinions and using that feedback to enhance offerings shows a commitment to continuous improvement.

Implementing loyalty programs is an effective way to incentivize repeat business. Rewarding clients for their loyalty with exclusive offers, discounts, or other benefits can reinforce their decision to stay with your brand.

Inevitably, challenges may arise, and how you handle them can make or break a client relationship. Addressing issues promptly and effectively, showing empathy and dedication to resolving problems, can turn a negative experience into a positive one, solidifying trust and loyalty.

Staying up-to-date with industry trends and advancements is essential to remain competitive. Clients value businesses that innovate and offer cutting-edge solutions, further establishing trust and reliability.

Word-of-mouth referrals are powerful marketing tools. Encourage clients to refer others to your business by providing exceptional service and offering referral rewards. Satisfied clients are more likely to share their positive experiences with others.

Building long-term client relationships and referral business requires consistent effort and dedication. It's not just about the initial sale but rather an ongoing commitment to providing value and exceptional experiences to clients.

Nurturing relationships with clients is an investment that pays off in the long run. Loyal clients are not only more likely to make repeat purchases but also tend to spend more than new customers. Additionally, loyal clients can become brand advocates, spreading positive word-of-mouth and referring potential new clients to your business.

One of the foundational elements of building long-term client relationships is delivering exceptional service. By going above and beyond to meet and exceed client expectations, businesses can create a positive and memorable experience for their clients. This could include providing high-quality products or services, offering personalized recommendations or solutions, and being attentive to clients' needs and preferences.

Trust is a fundamental aspect of any successful relationship, and the client-business relationship is no exception. Building trust with clients requires honesty, reliability, and consistency. Being transparent about your offerings, pricing, and policies, and delivering on your promises will help establish trustworthiness.

Personalization plays a critical role in fostering long-term client relationships. Clients want to feel valued and understood, and tailoring your products or services to match their specific requirements can go a long way in achieving this. Gathering data on client preferences and behavior can help you deliver personalized experiences that resonate with them.

Maintaining regular communication with clients is essential for staying top-of-mind and nurturing the relationship. This includes sending relevant updates, special offers, newsletters, or even personalized messages to show that you care about their needs and are there to support them.

Feedback from clients provides valuable insights into their

experiences with your business. Actively seeking feedback and using it to improve your offerings demonstrates that you value their opinions and are committed to continuous enhancement.

Implementing a well-designed loyalty program can incentivize clients to remain engaged with your business. Rewarding repeat purchases or referrals with exclusive perks, discounts, or loyalty points encourages clients to continue doing business with you.

Despite best efforts, occasional challenges or issues may arise. The key is to address these concerns promptly and effectively. Displaying genuine concern, offering solutions, and ensuring a positive resolution can turn a negative situation into an opportunity to strengthen the relationship.

Staying up-to-date with industry trends and developments is crucial for remaining competitive and relevant in the eyes of clients. Clients are more likely to stay loyal to a business that offers innovative and cutting-edge solutions.

Encouraging word-of-mouth referrals can be a powerful growth strategy. Delighted clients are more likely to recommend your business to their friends, family, or colleagues. Offering referral rewards or incentives can further motivate clients to share their positive experiences.

CHAPTER 12: SCALING AND GROWTH

12.1 Strategies for Scaling Your Construction Business

Scaling a construction business requires careful planning, strategic decision-making, and effective implementation of growth strategies. As the demand for construction services increases, businesses must adapt to meet the needs of a growing market while maintaining quality and efficiency. In this article, we will explore various strategies that can help construction companies scale their operations successfully.

One key strategy for scaling a construction business is to establish efficient project management processes. This involves implementing robust systems for estimating, scheduling, and resource allocation. By streamlining these processes, construction companies can optimize project timelines, reduce costs, and improve overall productivity. Leveraging project management software can also enhance communication and collaboration among team members, leading to smoother project execution.

Another critical aspect of scaling a construction business is building a strong team. Hiring and retaining skilled professionals who align with your company's values and vision is essential. As the business expands, it is vital to invest in training and development programs to upskill existing employees and attract top talent. A competent and motivated team is crucial for maintaining quality standards and meeting the demands of a growing customer base.

Effective networking and strategic partnerships play a significant role in scaling a construction business. Building relationships with suppliers, subcontractors, architects, and other industry professionals can open doors to new opportunities. Collaborating with trusted partners can help expand service offerings, access new markets, and share resources and expertise. Strategic alliances also enhance credibility and reputation in the industry, leading to increased client confidence and potential referrals.

Implementing technology and embracing innovation is a key driver of growth in the construction industry. Investing in construction-specific software and tools can streamline operations, improve accuracy, and enhance efficiency. For example, adopting Building Information Modeling (BIM) technology can improve project visualization, coordination, and reduce errors. Utilizing drones for site surveys and inspections can save time and provide accurate data. Embracing sustainable construction practices and green technologies not only contributes to environmental responsibility but can also attract environmentally conscious clients and projects.

Diversification of services is another effective strategy for scaling a construction business. By expanding into complementary sectors or offering additional services, companies can tap into new revenue streams. For example, a residential construction company can explore commercial or industrial projects. Adding services such as renovation, maintenance, or property management can also create opportunities for recurring revenue and long-term client relationships.

Marketing and branding efforts play a crucial role in scaling a construction business. Developing a strong brand identity, highlighting unique selling propositions, and effectively communicating the company's value proposition can help attract new clients and differentiate from competitors. Leveraging digital marketing channels, such as social media, content marketing, and search engine optimization, can expand the company's online presence and generate leads. Utilizing client testimonials, case studies, and project portfolios can also demonstrate expertise and build trust with potential clients.

Access to capital and financial management are fundamental considerations when scaling a construction business. Adequate funding is often required to invest in new equipment, technology, talent, and marketing efforts. Exploring financing options such as loans, lines of credit, or partnerships with investors can provide the necessary capital for growth. Effective financial management, including accurate accounting practices, budgeting, and monitoring key performance indicators, is essential to ensure sustainable growth and profitability.

Lastly, maintaining a focus on quality, safety, and customer satisfaction is paramount. As the business expands, it is crucial to uphold high standards and deliver exceptional results. Prioritizing quality craftsmanship, adhering to safety protocols, and providing excellent customer service will help build a strong reputation and foster long-term client relationships.

12.2 Expanding into New Markets and Services

Expanding into new markets and services is a crucial step for businesses looking to grow and diversify their operations. It offers opportunities to tap into untapped customer segments, increase revenue streams, and mitigate risks associated with relying solely on existing markets. However, venturing into new territories or introducing new services requires careful planning, market research, and strategic execution.

Before expanding into new markets, it is essential to conduct thorough market research. This includes analyzing the target market's size, potential demand, competition, regulatory environment, cultural nuances, and customer preferences. This information will help businesses make informed decisions and tailor their offerings to suit the specific needs and preferences of the new market.

Identifying the right entry strategy is crucial when expanding into new markets. Companies can choose to establish a physical presence through setting up offices or retail outlets, form strategic partnerships with local businesses, or enter the market through e-commerce platforms. The chosen strategy should align with the company's resources, capabilities, and long-term objectives.

Adapting products or services to meet the unique needs and preferences of the new market is essential for success. Localization efforts may include modifying product features, packaging, pricing, and branding to resonate with the target market. This demonstrates an understanding of the local culture and builds trust with potential customers.

Building a strong distribution network is vital when expanding into new markets. Establishing relationships with local distributors, wholesalers, or retailers can help reach customers more effectively. Collaborating with local partners who have a deep understanding of the market can also provide valuable insights and facilitate market penetration.

Effective marketing and promotion play a significant role in introducing products or services to a new market. Developing a comprehensive marketing strategy that includes a mix of traditional and digital marketing channels can help raise awareness, generate leads, and build a customer base. Localized marketing campaigns that consider cultural nuances and preferences can be more impactful in capturing the attention of the target audience.

Expanding into new services requires a similar level of planning and research. Assessing market demand, competition, and profitability of potential new services is crucial. It is important to evaluate if the existing infrastructure, resources, and expertise of the business align with the requirements of the new services or if additional investment or training is necessary.

Diversifying services can offer several advantages, such as cross-selling opportunities to existing customers and attracting new customers who may have different needs. However, it is essential to ensure that the expansion into new services is aligned with the core competencies of the business and supports the overall business strategy.

Developing a clear value proposition and differentiating the new services from competitors is critical. Highlighting unique features, benefits, or expertise can help position the business as a preferred choice in the market. Testimonials, case studies, and references from existing clients can also instill confidence in potential customers regarding the quality and reliability of the new services.

When expanding into new markets or introducing new services, businesses should be prepared to invest time, resources, and effort in building brand awareness and establishing a reputation. It may take time to gain traction and see a return on investment, so patience and perseverance are key.

Regular monitoring and evaluation of performance metrics are necessary to assess the success of the expansion efforts. This includes tracking sales, customer feedback, market share, and profitability. Making data-driven decisions and adapting strategies based on market response will help optimize performance and drive growth in new markets or services.

12.3 Sustainable Growth and Long-Term Success

Sustainable growth and long-term success are fundamental goals for businesses seeking stability and prosperity. Achieving sustainable growth involves striking a balance between expansion, profitability, and responsible practices that consider social, environmental, and economic factors. This article will explore various aspects of sustainable growth and how businesses can adopt strategies to ensure their long-term success.

One of the essential elements of sustainable growth is financial stability. Businesses must maintain a healthy balance between revenue generation and expenses, ensuring that they can cover operational costs and reinvest in growth. Proper financial management, including budgeting, forecasting, and risk assessment, is crucial to avoid overextension and potential financial pitfalls.

In addition to financial stability, a focus on innovation is vital for long-term success. Embracing innovation allows businesses to adapt to changing market dynamics, stay ahead of competitors, and meet evolving customer demands. This may involve investing in research and development, adopting new technologies, or introducing innovative products and services.

Customer-centricity plays a significant role in sustainable growth and long-term success. Understanding customer needs, preferences, and pain points enables businesses to develop tailored solutions and build lasting relationships. Repeat business and word-of-mouth referrals from satisfied customers are critical drivers of sustainable growth.

Another key aspect of sustainable growth is maintaining a strong corporate culture and employee engagement. A positive work environment that fosters employee satisfaction, professional growth, and work-life balance contributes to productivity and reduces turnover. Engaged employees are more likely to be committed to the company's mission and contribute to its long-term success.

Emphasizing social and environmental responsibility is increasingly becoming a critical factor for sustainable growth. Businesses that demonstrate commitment to ethical practices, environmental sustainability, and social impact tend to attract customers, investors, and partners who align with these values. Adopting sustainable practices not only benefits the planet but also contributes to brand reputation and customer loyalty.

Strategic planning and flexibility are essential for businesses to navigate uncertainties and challenges successfully. Sustainable growth requires businesses to set clear long-term goals and develop actionable plans to achieve them. However, being flexible and adaptable allows businesses to adjust their strategies in response to changing market conditions or unforeseen events.

Fostering partnerships and collaborations can accelerate sustainable growth. Strategic alliances with like-minded businesses or organizations can lead to shared resources, expanded networks, and access to new markets or technologies. Collaborative efforts often leverage complementary strengths, boosting competitiveness and innovation.

Incorporating scalability into business operations is crucial for long-term success. As a business grows, it must be capable of handling increased demand without compromising the quality of its products or services. Implementing scalable processes, technology, and infrastructure prepares a business for future growth.

Continuous learning and improvement are essential for sustainable growth. Monitoring industry trends, customer feedback, and competitor activities can inform business decisions and drive innovation. Businesses that prioritize learning and seek opportunities for improvement are better positioned to adapt to market changes and maintain relevance.

Investing in human capital is a critical aspect of sustainable growth. Hiring and retaining talented and skilled employees are instrumental in achieving long-term success. Businesses should offer ongoing training and development opportunities to enhance employee skills and knowledge, empowering them to contribute effectively to the organization's growth.

Staying true to a strong set of core values and a clear vision is essential for sustainable growth. These guiding principles provide a framework for decision-making and ensure consistency across the organization. Aligning actions with values fosters trust among stakeholders and supports the company's long-term objectives.

Measuring and tracking key performance indicators (KPIs) is essential for assessing progress toward sustainable growth goals. KPIs can vary depending on the business's industry and objectives but may include financial metrics, customer satisfaction scores, employee retention rates, and environmental impact measurements.

Businesses that prioritize sustainable growth and commit to responsible practices are more likely to thrive in a competitive landscape, build lasting relationships with customers and stakeholders, and contribute positively to the world around them.

12.4 Staying Competitive in the Construction Industry

Staying competitive in the construction industry is a constant challenge that requires adaptability, innovation, and a focus on efficiency. With advancements in technology, changes in customer preferences, and evolving market dynamics, construction companies must continuously seek ways to differentiate themselves and deliver value to their clients.

One of the key strategies for staying competitive is embracing technology and innovation. Construction firms can leverage technologies such as Building Information Modeling (BIM), drones, 3D printing, and virtual reality to enhance project visualization, streamline communication, and improve project efficiency. Adopting these tools not only improves project outcomes but also positions companies as forward-thinking and attractive to clients seeking cutting-edge solutions.

Moreover, construction companies must prioritize sustainability and green practices to remain competitive. Sustainable construction methods and materials are becoming increasingly important to clients who value environmentally responsible projects. Integrating sustainability into construction practices not only helps the environment but also opens up opportunities for businesses to bid on eco-friendly projects and win contracts.

Emphasizing safety is another critical aspect of competitiveness in the construction industry. Companies that prioritize safety on their worksites can reduce accidents, avoid costly delays, and build a reputation as a reliable and responsible contractor. Implementing comprehensive safety training and maintaining strict adherence to safety protocols demonstrates a commitment to the well-being of workers and clients alike.

In a highly competitive industry, reputation and brand image play a significant role in attracting new clients. Companies must consistently deliver high-quality work, meet deadlines, and provide excellent customer service to build a positive reputation. Satisfied clients are more likely to recommend the construction company to others, contributing to word-of-mouth referrals, which are powerful drivers of business growth.

Offering competitive pricing while maintaining profitability is a balancing act that construction companies must master. Accurate cost estimation, efficient project management, and smart procurement practices are essential for controlling costs without compromising on quality. Striking the right balance between competitive pricing and profit margin ensures a sustainable business model.

Forming strategic partnerships and collaborations can also enhance competitiveness in the construction industry. Collaborating with architects, engineers, suppliers, or other construction firms can expand the range of services offered and provide access to new markets. These alliances can create synergies that lead to greater efficiency and innovation.

Investing in the continuous training and development of employees is essential for staying competitive. Skilled and knowledgeable staff contribute to better project execution, higher productivity, and improved customer satisfaction. Encouraging employee growth and providing opportunities for professional development also help retain talent and build a strong team.

Incorporating Lean construction principles can be a game-changer for companies aiming to be more competitive. Lean methodologies focus on eliminating waste, optimizing processes, and maximizing value for the client. Adopting Lean practices can result in improved productivity, reduced project timelines, and better resource management.

Maintaining a strong financial position is critical for surviving economic downturns and seizing opportunities for growth. Construction companies should prioritize financial stability, manage cash flow effectively, and be prepared to weather unforeseen challenges. This financial strength can provide a competitive advantage in a volatile market.

Building strong relationships with clients and understanding their unique needs is vital for staying competitive. By delivering personalized solutions and excellent customer service, construction companies can create loyal clients who trust them for future projects. Repeat business and long-term relationships contribute to a stable revenue stream and a competitive edge.

Diversification can also enhance competitiveness in the construction industry. Exploring new markets, such as residential, commercial, industrial, or infrastructure projects, can provide opportunities for growth and reduce dependence on a single sector. Offering additional services, such as renovation, maintenance, or sustainable construction, can also broaden the company's offerings and attract a diverse clientele.

Investing in marketing and branding efforts is essential for increasing visibility and attracting new clients. Effective marketing strategies, both online and offline, can help construction companies showcase their expertise, highlight successful projects, and demonstrate their commitment to quality. An impactful brand image can differentiate a company in a crowded marketplace.

Staying competitive in the construction industry requires a multi-faceted approach that involves embracing technology and innovation, prioritizing sustainability and safety, building a positive reputation, offering competitive pricing, forming strategic partnerships, investing in employee development, adopting Lean practices, maintaining financial stability, focusing on client relationships, diversifying services, and investing in marketing and branding. By continually evolving and adapting to market trends and customer demands, construction companies can position themselves for success, secure new business opportunities, and achieve sustainable growth in a dynamic and competitive industry.

CHAPTER 13: CASE STUDIES AND SUCCESS STORIES

13.1 Real-Life Examples of Successful Construction Businesses

C ase studies and success stories are powerful tools that showcase real-world examples of how businesses have overcome challenges, achieved their goals, and delivered value to their clients. These narratives provide valuable insights into a company's capabilities, innovative approaches, and the positive impact of their products or services. By presenting specific situations and outcomes, case studies and success stories offer credibility and build trust with potential customers, investors, and partners. In this article, we will explore the significance of case studies and success stories and how they contribute to business growth and success.

Case studies are detailed examinations of specific projects, situations, or client interactions that a business has encountered. They often highlight the unique challenges faced, the strategies employed, and the results achieved. Case studies can cover various aspects of a business's operations, such as successful product launches, complex problem-solving, cost-effective solutions, or the implementation of new technologies. By presenting the context, actions, and outcomes in a structured format, case studies provide valuable insights into a company's expertise and the value it brings to clients.

Success stories, on the other hand, focus on the positive outcomes and achievements of a business or its clients. They highlight how a product or service has made a significant difference in a customer's life or business operations. Success stories often showcase quantifiable results, such as increased revenue, improved efficiency, or enhanced customer satisfaction. These narratives evoke positive emotions and build brand affinity, encouraging potential customers to envision the benefits they could also experience.

Both case studies and success stories are powerful marketing tools. They allow businesses to demonstrate their expertise, industry knowledge, and problem-solving capabilities in a tangible and relatable manner. By presenting evidence of successful projects or satisfied customers, businesses can build credibility and trust, which are essential for attracting new clients and retaining existing ones.

Case studies and success stories provide a human touch to a company's brand. By sharing the experiences of real clients or customers, businesses create relatable and authentic narratives. Prospective clients can see themselves in similar situations and find confidence in the company's ability to deliver positive outcomes. These stories create emotional connections that go beyond typical marketing messages.

From a sales perspective, case studies and success stories act as persuasive tools. They help sales teams address objections and provide evidence of a company's capabilities and track record. Sharing relevant case studies during the sales process can help overcome client skepticism and demonstrate the company's suitability for a particular project or need.

In addition to sales, case studies and success stories play a crucial role in business development and relationship building. When approaching potential partners, investors, or collaborators, sharing success stories showcases a business's accomplishments and creates a favorable impression. It can also open doors to new opportunities and foster strategic alliances.

Publishing case studies and success stories on a company's website or marketing materials is an effective inbound marketing strategy. When potential clients conduct online research, these narratives provide valuable content that educates and informs, making the company more visible and engaging. They can be optimized for search engines to attract relevant traffic and generate leads.

For B2B companies, case studies and success stories are particularly valuable. In complex purchasing decisions, decision-makers often seek evidence of a vendor's capability to deliver on promises. Case studies provide concrete proof that a business can address specific challenges and deliver results.

Creating compelling case studies and success stories requires a thoughtful approach. Companies must ensure that they have permission from clients or customers to share their experiences publicly. It is crucial to gather accurate data and metrics that demonstrate the impact of the company's products or services. Presenting information in a clear and concise format, with a focus on the client's journey and the achieved results, enhances the effectiveness of these narratives.

13.2 Lessons Learned from Case Studies and Success Stories

Lessons learned from case studies and success stories offer valuable insights that businesses can apply to their own operations, strategies, and decision-making processes. By analyzing these real-world examples of triumphs and challenges, companies can gain a deeper understanding of what works, what doesn't, and how to navigate obstacles to achieve success. In this article, we will explore some of the key lessons that can be drawn from case studies and success stories and how businesses can leverage this knowledge to enhance their performance and growth.

Customer-Centricity

One common theme in many case studies and success stories is the focus on customer-centricity. Successful businesses prioritize understanding their customers' needs, pain points, and preferences. They tailor their products or services to address these specific requirements, leading to higher customer satisfaction and loyalty. Lessons from these stories emphasize the importance of listening to customers, gathering feedback, and continuously improving to meet evolving expectations.

Innovation and Adaptability

Case studies and success stories often feature companies that embrace innovation and adaptability. They are willing to explore new technologies, processes, and business models to stay ahead of the competition. These stories emphasize the need for businesses to be agile and open to change, as innovation is a driving force for growth and long-term success.

Strategic Planning

The success of many businesses showcased in case studies can be attributed to effective strategic planning. Having a clear vision, setting achievable goals, and developing actionable plans are critical for business success. Strategic planning helps businesses align their efforts, allocate resources wisely, and make informed decisions that drive progress.

Problem-Solving and Resilience

Case studies often feature businesses that have encountered challenges and obstacles. The key to their success lies in their ability to tackle these problems with creativity, resilience, and determination. Lessons from these stories highlight the importance of problem-solving skills and the willingness to persevere in the face of adversity.

Building Strong Teams

The impact of a motivated and skilled team is evident in many success stories. Companies that invest in hiring and retaining top talent and foster a positive work culture tend to achieve better outcomes. These stories emphasize the importance of building strong teams that collaborate effectively and share a common vision.

Risk Management

Risk management is a critical aspect of business success, as highlighted in case studies. Companies that assess and mitigate risks proactively are better equipped to navigate uncertainties and safeguard their operations. Understanding potential risks allows businesses to develop contingency plans and make informed decisions.

Continuous Learning and Improvement

Many successful businesses are committed to continuous learning and improvement. They invest in employee development, stay informed about industry trends, and adapt their strategies accordingly. These stories underscore the value of being open to learning, analyzing performance data, and making data-driven improvements.

Transparency and Communication

Transparency and effective communication are essential components of successful businesses. In case studies, companies that prioritize open communication with stakeholders, including clients, employees, and partners, tend to build trust and foster strong relationships. These stories emphasize the need for clear and transparent communication to avoid misunderstandings and ensure alignment with all parties involved.

Sustainability and Social Responsibility

Case studies and success stories increasingly highlight the importance of sustainability and social responsibility. Businesses that integrate ethical practices, environmental consciousness, and social impact into their operations tend to resonate better with customers and gain a competitive advantage. These stories underline the significance of responsible business practices in today's market.

Flexibility in a Changing Market
Successful businesses demonstrate flexibility and adaptability to changing market conditions. They are attuned to customer demands, technological advancements, and industry shifts. Lessons from these stories emphasize the need to remain flexible and responsive to maintain relevance and competitiveness.

13.3 Inspirational Stories of Overcoming Challenges in the Construction Industry

The construction industry is no stranger to challenges, and inspirational stories of overcoming these hurdles showcase the resilience, determination, and ingenuity of professionals in this field. These stories highlight how individuals and companies have triumphed over adversity, turning setbacks into opportunities for growth and success. In this article, we will explore some inspirational stories from the construction industry that demonstrate the power of perseverance, innovation, and teamwork.

One of the most common challenges in the construction industry is completing projects within tight deadlines. In an inspiring story, a construction team faced numerous delays due to unforeseen weather conditions and material shortages. Instead of succumbing to frustration, the team came together to devise creative solutions. They reorganized the project schedule, implemented 24/7 work shifts, and collaborated closely with suppliers to expedite material deliveries. Through their collective efforts and determination, the project was completed on time, earning praise from the client and solidifying the team's reputation for meeting challenges head-on.

Safety is paramount in the construction industry, and overcoming safety-related challenges requires unwavering commitment. An inspiring story involves a construction company that had a history of safety incidents and accidents on their worksites. Recognizing the need for a transformative change in safety culture, the company invested heavily in safety training and awareness programs. They engaged their employees in safety initiatives and provided incentives for adhering to safety protocols. Over time, their efforts paid off, and the company saw a significant reduction in accidents and injuries. This transformation not only improved the well-being of their workforce but also positively impacted their reputation in the industry.

Financial constraints are another common challenge in construction projects. In one inspirational story, a small construction firm secured a major contract to build a community center. However, they faced difficulties securing funding to cover the project's upfront costs. Determined not to let this opportunity slip away, they explored alternative financing options, including crowdfunding and community partnerships. Their resourcefulness paid off, and they successfully raised the necessary funds to start the project. The community rallied behind the construction team, and the completed community center became a symbol of collective effort and determination.

The construction industry often encounters unexpected technical challenges that require innovative solutions. In a notable story, an engineering firm faced a complex engineering problem while designing a bridge in a seismic zone. The conventional design approaches did not adequately address the specific geological conditions. Instead of giving up, the engineering team collaborated with geologists and researchers to develop a groundbreaking design that could withstand seismic forces effectively. This innovative approach not only solved the engineering challenge but also earned the firm recognition for their ingenuity and problem-solving skills.

Inspirational stories also demonstrate the power of perseverance in large-scale construction projects. One such story involves the construction of an iconic skyscraper. The project faced numerous setbacks, including adverse weather, material shortages, and labor disputes. Despite these challenges, the construction team persisted with unwavering determination. They worked around the clock, employed state-of-the-art technology, and implemented strict project management to keep the project on track. The completed skyscraper became a symbol of human determination and engineering marvel, standing tall as a testament to overcoming adversity.

Innovation and technological advancements have also played a significant role in overcoming challenges in the construction industry. For instance, a construction company facing a labor shortage leveraged robotic construction technology to increase productivity and efficiency. The use of robots not only addressed the labor shortage but also improved overall construction quality and safety. This inspirational story illustrates how embracing technology can lead to transformative improvements in the construction process.

Collaboration and teamwork are essential ingredients for overcoming challenges in the construction industry. In one uplifting story, multiple construction companies collaborated on a complex infrastructure project. Each company brought its unique expertise to the table, and their collective effort resulted in a seamless execution. This collaboration not only delivered exceptional results but also fostered a spirit of camaraderie among the teams, setting a positive example for future joint ventures.

Inspirational stories in the construction industry are not limited to large-scale projects. Even small construction businesses face challenges that demand resourcefulness and determination. For example, a local construction company faced tough competition from larger firms in the area. To stand out, they focused on providing exceptional customer service, personalized solutions, and competitive pricing. Over time, their dedication to client satisfaction earned them a loyal customer base and referrals, solidifying their position in the market.

The construction industry is ever-evolving, and inspirational stories of overcoming challenges continue to emerge. Whether it is adapting to new technologies, navigating regulatory changes, or addressing environmental sustainability, these stories inspire the entire construction community to persevere and innovate. They remind us that challenges are not roadblocks but stepping stones toward growth and success.

Inspirational stories of overcoming challenges in the construction industry highlight the unwavering spirit, creativity, and dedication of professionals in this field. From completing projects within tight deadlines to transforming safety cultures, finding innovative solutions to technical problems, and leveraging technology for efficiency, these stories showcase the resilience of construction teams. They underscore the significance of collaboration, perseverance, and customer-centricity in achieving success. By learning from these stories, construction professionals can draw inspiration and apply valuable lessons to navigate challenges and drive their businesses forward.

CHAPTER 14: CONCLUSION

14.1 Recap of Key Points to Know Before Starting a Construction Business

Starting a construction business requires careful planning, preparation, and a deep understanding of the industry. Before embarking on this journey, it is crucial to conduct thorough market research to assess the demand for construction services in your target area and identify potential competitors. This will help you determine the viability and potential for growth in your chosen market.

Developing a comprehensive business plan is essential. This plan should outline your business goals, target market, services offered, marketing strategies, and financial projections. It serves as a roadmap for your business and helps guide your decision-making process.

Complying with legal and regulatory requirements is critical when starting a construction business. You need to research and understand the licenses, permits, insurance, and safety regulations specific to your location. Ensuring compliance will protect your business legally and help you maintain a good reputation in the industry.

Securing adequate funding is a key consideration. You need to determine the startup costs, including equipment, tools, vehicles, and initial marketing expenses. Explore different funding options such as personal savings, loans, or investment partnerships. Proper financial management, including tracking expenses, managing cash flow, and monitoring profitability, is vital for the long-term success of your business.

Specialization and expertise play a crucial role in standing out in the construction industry. Determine the specific construction services you will offer based on your expertise, resources, and market demand. Focusing on a niche area can help you develop a competitive advantage and differentiate yourself from broader service providers.

Building a skilled and reliable team is essential. Hire employees who have the necessary skills and experience to carry out construction projects efficiently. Additionally, invest in ongoing training and professional development to enhance their skills and keep them up to date with industry trends.

Networking and forming strategic partnerships are key to growing your construction business. Attend industry events, join professional associations, and establish relationships with suppliers, subcontractors, architects, and other professionals in the construction industry. These connections can lead to collaboration opportunities, referrals, and access to new projects.

Marketing and promoting your construction business are crucial for attracting clients. Develop a strong brand identity and create a professional website that showcases your services, previous projects, and client testimonials. Utilize digital marketing strategies such as search engine optimization (SEO), social media marketing, and online advertising to reach your target audience effectively.

Maintaining a focus on quality and customer satisfaction is paramount. Delivering high-quality work, meeting project deadlines, and providing exceptional customer service will help you build a positive reputation and foster long-term client relationships. Satisfied clients are more likely to recommend your services and contribute to the growth of your business.

Finally, staying informed about industry trends, technological advancements, and best practices is crucial. Continuously educate yourself and adapt to changes in the construction industry. This will help you stay competitive, deliver innovative solutions, and position your business for long-term success.

14.2 Encouragement and Actionable Steps for Aspiring Construction Entrepreneurs

Gaining industry experience is crucial for aspiring construction entrepreneurs. Before starting your own business, spend time working in the construction industry to learn the ins and outs of the trade. This hands-on experience will give you valuable insights into the construction process, project management, client interactions, and the challenges that may arise on construction sites. Working alongside experienced professionals will also allow you to build a network of contacts, which can be beneficial when starting your own venture.

Education and continuous learning play a significant role in the construction industry. Consider pursuing relevant certifications, diplomas, or degrees in construction management, engineering, or related fields. These educational qualifications will not only enhance your knowledge but also boost your credibility as a construction entrepreneur. Stay updated with industry trends, best practices, and technological advancements to remain competitive in the market.

Developing a solid business plan is essential for any aspiring entrepreneur. A well-crafted business plan should outline your business goals, target market, services offered, marketing strategies, financial projections, and plans for growth. It acts as a roadmap for your construction business and helps you stay focused on your objectives.

Understanding the legal and regulatory requirements for starting a construction business is crucial. Depending on your location, you may need to obtain licenses, permits, and insurance to operate legally. Complying with safety regulations is also vital to protect your workers, clients, and your business from potential liabilities.

Networking and building relationships are integral to success in the construction industry. Attend industry events, join professional associations, and engage with potential clients, suppliers, subcontractors, and other industry professionals. These connections can lead to collaborative opportunities, referrals, and valuable insights.

Securing funding is often a significant challenge for aspiring construction entrepreneurs. Explore various funding options, such as personal savings, loans, or seeking investors. Develop a realistic budget that accounts for startup costs, equipment, materials, marketing, and ongoing expenses. Proper financial management is crucial to ensure the financial stability of your construction business.

Building a strong team is essential for the success of any construction business. Hire skilled and motivated individuals who share your vision and values. Invest in training and development to enhance their skills and foster a positive work culture. A competent and dedicated team will contribute to the growth and reputation of your business.

Marketing your construction business is critical to attracting clients and building a brand presence. Create a professional website that showcases your services, previous projects, and client testimonials. Utilize digital marketing strategies such as search engine optimization (SEO), social media marketing, and online advertising to reach your target audience effectively.

Offering exceptional customer service is a surefire way to

stand out in the competitive construction industry. Building strong relationships with clients, understanding their needs, and delivering high-quality work will lead to repeat business and positive word-of-mouth referrals.

Embrace innovation and technology to stay ahead of the competition. Explore construction software, Building Information Modeling (BIM), drones, and other digital tools that can enhance project efficiency and communication. Embracing technology can improve productivity and set your construction business apart from traditional approaches.

Take calculated risks and be open to learning from failures. As an entrepreneur, you may encounter challenges and setbacks along the way. Learn from these experiences and use them as opportunities to grow and improve. Stay resilient and maintain a positive mindset, even in the face of adversity.

BOOKS BY THIS AUTHOR

The Dictionary Of Construction Terminologies: A Compendium Of Knowledge For Students, Academics, Practitioners, And House Owners Paperback

The dictionary of construction terminologies book is a comprehensive reference guide that provides definitions and explanations of the technical language and jargon used in the construction industry. It is an invaluable resource for professionals working in construction, as well as for students learning about the industry or for individuals looking to understand construction-related concepts better.

The book features a wide range of entries that cover various aspects of construction, including architecture, engineering, materials, equipment, and techniques. The book also provides clear and concise definitions of technical terms, written in easy-to-understand language. Terminologies are presented alphabetically to help readers find the descriptions they need quickly and easily. Whether you are a professional working in the field or interested in construction, this book is an essential tool to help you navigate the complex world of construction terminology with confidence and clarity.

Are you a student of construction, a house owner, an academic in the construction industry, or a practitioner that desires to acquire more knowledge about construction terms? If your answer to the preceding question is affirmative, this book may be one of the best investments you will ever make.

Construction Health And Safety Fundamentals

In the fast-paced world of construction, ensuring the health and safety of workers is paramount. This book serves as an effective guide, laying the foundation for creating a culture of safety and excellence in the construction industry.

Offering a comprehensive exploration of fundamental principles, strategies, and best practices, this book covers a wide range of topics. From safe operation of heavy machinery to crane and hoist safety, rigging and lifting operations, power tool safety, electrical safety, fire safety and emergency response, working at heights and fall protection, confined space entry and rescue, and much more.

Each chapter provides clear explanations of key concepts and actionable insights. One distinguishing feature of the book is its comprehensive approach to emerging trends and technologies in construction health and safety. It explores innovative solutions such as wearable technologies, virtual reality training, and predictive analytics, empowering readers to stay ahead of the curve and leverage cutting-edge tools for enhanced safety outcomes.

www.ingramcontent.com/pod-product-compliance
Lightning Source LLC
Chambersburg PA
CBHW072213290526
45794CB00004B/1738